WATER
NO LONGER TAKEN
FOR GRANTED

ISSN 1536-5212

WATER
NO LONGER TAKEN FOR GRANTED

Sandra M. Alters

INFORMATION PLUS® REFERENCE SERIES
Formerly Published by Information Plus, Wylie, Texas

THOMSON
™
GALE

Detroit • New York • San Francisco • New Haven, Conn. • Waterville, Maine • London

Water: No Longer Taken for Granted

Sandra M. Alters

Paula Kepos, Series Editor

Project Editors
Kathleen J. Edgar, John McCoy

Permissions
Aja Perales, Jhanay Williams

Composition and Electronic Prepress
Evi Seoud

Manufacturing
Cynde Bishop

ISBN-13: 978-0-7876-5103-9 (set)
ISBN-10: 0-7876-5103-6 (set)
ISBN-13: 978-1-4144-0780-7
ISBN-10: 1-4144-0780-7
ISSN 1536-5212

This title is also available as an e-book.
ISBN-13: 978-1-4144-2948-9 (set), ISBN-10: 1-4144-2948-7 (set)
Contact your Gale Group sales representative for ordering information.

Printed in the United States of America
10 9 8 7 6 5 4 3 2 1

TABLE OF CONTENTS

PREFACE

Water: No Longer Taken for Granted is part of the *Information Plus Reference Series*. The purpose of each volume of the series is to present the latest facts on a topic of pressing concern in modern American life. These topics include today's most controversial and most studied social issues: abortion, capital punishment, care for senior citizens, crime, the environment, health care, immigration, minorities, national security, social welfare, women, youth, and many more. Although written especially for the high school and undergraduate student, this series is an excellent resource for anyone in need of factual information on current affairs.

By presenting the facts, it is the Gale Group's intention to provide its readers with everything they need to reach an informed opinion on current issues. To that end, there is a particular emphasis in this series on the presentation of scientific studies, surveys, and statistics. These data are generally presented in the form of tables, charts, and other graphics placed within the text of each book. Every graphic is directly referred to and carefully explained in the text. The source of each graphic is presented within the graphic itself. The data used in these graphics are drawn from the most reputable and reliable sources, in particular from the various branches of the U.S. government and from major independent polling organizations. Every effort has been made to secure the most recent information available. The reader should bear in mind that many major studies take years to conduct and that additional years often pass before the data from these studies are made available to the public. Therefore, in many cases the most recent information available in 2007 dated from 2004 or 2005. Older statistics are sometimes presented as well, if they are of particular interest and no more-recent information exists.

Although statistics are a major focus of the *Information Plus Reference Series*, they are by no means its only content. Each book also presents the widely held positions and important ideas that shape how the book's subject is discussed in the United States. These positions are explained in detail and, where possible, in the words of their proponents. Some of the other material to be found in these books includes: historical background; descriptions of major events related to the subject; relevant laws and court cases; and examples of how these issues play out in American life. Some books also feature primary documents or have pro and con debate sections giving the words and opinions of prominent Americans on both sides of a controversial topic. All material is presented in an even-handed and unbiased manner; the reader will never be encouraged to accept one view of an issue over another.

HOW TO USE THIS BOOK

Water is one of the most vital resources on Earth. Most of the human body consists of water, and any person who is deprived of water for a significant period will die. The same can be said of all the animals and plants that people rely on for food. Only the air we breathe is more essential to human life, but while there is no shortage of air around the world, there are many regions where drinkable water is in short supply, such as the western United States. Furthermore, even where water is available, human activities may have contaminated it with diseases and chemicals, with serious consequences for the people and wildlife that use or live in it. As a result, the use and care of the water that is available is a controversial topic in the United States and around the world. This book explores all of the major issues surrounding water in America, from water usage rights to pollution control to safe drinking water.

Water: No Longer Taken for Granted consists of eight chapters and three appendixes. Each of the chapters is devoted to a particular aspect of water in the United States. For a summary of the information covered in each chapter, please see the synopses provided in the Table of Contents at the front of the book. Chapters generally begin with an overview of the basic facts and background information on the chapter's topic, then proceed to examine subtopics of particular interest. For example, Chapter 5, Drinking Water—Safety on Tap, begins with a description of how much water Americans use, then discusses the water supply system in the United States, both public and private. This is followed by a detailed discussion of the contaminants most often found in drinking water, including chemicals such as arsenic and microbes such as coliform bacteria. The emphasis is on their possible sources and effects. The chapter then moves on to describe the modern water treatment system, and how it removes these contaminants and otherwise makes water fit to drink. The fluoridation of water is also discussed. Next, the chapter describes the many laws that regulate the safety of drinking water, as well as the cost of safe water. Another major section is devoted to analyses of just how safe U.S. tap water is. The chapter concludes with a number of sections on issues of special interest or concern, such as protecting drinking water supplies, bottled water use in the United States, and drinking water in other countries. Readers can find their way through a chapter by looking for the section and subsection headings, which are clearly set off from the text. Or, they can also refer to the book's extensive Index, if they already know what they are looking for.

Statistical Information

The tables and figures featured throughout *Water: No Longer Taken for Granted* will be of particular use to the reader in learning about this topic. These tables and figures represent an extensive collection of the most recent and valuable statistics on water and related issues—for example, graphics in the book cover distribution of the world's water; groundwater use in the United States; the number and types of waterborne-disease outbreaks associated with recreational water use; the types of drinking water systems; industrial water use by state; and what percentage of the public is concerned about the pollution of drinking water. The Gale Group believes that making this information available to the reader is the most important way in which we fulfill the goal of this book: to help readers understand the issues and controversies surrounding water in the United States and to reach their own conclusions.

Each table or figure has a unique identifier appearing above it for ease of identification and reference. Titles for the tables and figures explain their purpose. At the end of each table or figure, the original source of the data is provided.

In order to help readers understand these often complicated statistics, all tables and figures are explained in the text. References in the text direct the reader to the relevant statistics. Furthermore, the contents of all tables and figures are fully indexed. Please see the opening section of the Index at the back of this volume for a description of how to find tables and figures within it.

Appendixes

In addition to the main body text and images, *Water: No Longer Taken for Granted* has three appendixes. The first is the Important Names and Addresses directory. Here the reader will find contact information for a number of government and private organizations that can provide further information on water. The second appendix is the Resources section, which can also assist the reader in conducting his or her own research. In this section, the author and editors of *Water: No Longer Taken for Granted* describe some of the sources that were most useful during the compilation of this book. The final appendix is the Index.

ADVISORY BOARD CONTRIBUTIONS

The staff of Information Plus would like to extend its heartfelt appreciation to the Information Plus Advisory Board. This dedicated group of media professionals provides feedback on the series on an ongoing basis. Their comments allow the editorial staff who work on the project to make the series better and more user-friendly. Our top priorities are to produce the highest-quality and most useful books possible, and the Advisory Board's contributions to this process are invaluable.

The members of the Information Plus Advisory Board are:

- Kathleen R. Bonn, Librarian, Newbury Park High School, Newbury Park, California

- Madelyn Garner, Librarian, San Jacinto College–North Campus, Houston, Texas

- Anne Oxenrider, Media Specialist, Dundee High School, Dundee, Michigan

- Charles R. Rodgers, Director of Libraries, Pasco-Hernando Community College, Dade City, Florida

- James N. Zitzelsberger, Library Media Department Chairman, Oshkosh West High School, Oshkosh, Wisconsin

COMMENTS AND SUGGESTIONS

The editors of the *Information Plus Reference Series* welcome your feedback on *Water: No Longer Taken for Granted*. Please direct all correspondence to:

Editors
Information Plus Reference Series
27500 Drake Rd.
Farmington Hills, MI 48331-3535

CHAPTER 1
WHAT IS WATER?

Most people living in the United States assume they will have plenty of clean, safe water for drinking, that crops and gardens can be regularly irrigated, and that sewage will be taken care of by their local treatment plant. In many parts of the world, however, the availability of water for personal and public use cannot be taken for granted. In fact, according to the World Health Organization (WHO), in "Water, Sanitation, and Hygiene Links to Health: Facts and Figures" (November 2004, http://www.who.int/water _ sanitation_health/factsfigures2005.pdf), 17% of the population around the world—1.1 billion people—did not have access to safe water in 2002.

Water is vital to human survival. Although people can survive for a month—possibly two—without food, they would die in about a week without water. Water is the most common substance on Earth. It covers three-fourths of the earth's surface and makes up about 65% of the adult human body, including 90% of its blood and 75% of its brain. Water is the main ingredient in most of the fruits, vegetables, and meats that people eat. According to the fact sheet "Safe Drinking Water Act 30th Anniversary: Water Facts" (June 2004, http://www.epa .gov/ogwdw/sdwa/30th/factsheets/waterfacts.html) by the U.S. Environmental Protection Agency, water makes up about 75% of a chicken, 80% of a pineapple, and 95% of a tomato. In *Water—More Nutrition per Drop* (2004, http://www.siwi.org/downloads/More_Nutrition_Per_Drop .pdf), the Stockholm International Water Institute and the International Water Management Institute report that growing enough food to produce an adequate diet for a human being for one year requires 1,300 meters cubed (343,421 gallons) per person per year, or about 941 gallons of water per person per day.

Even though it is essential to human existence, water can cause severe damage and destruction. It is terror to the swimmer caught up in a current. It contributes to rusting in cars and rotting in wood. Water in the form of hailstones can destroy crops, and in the winter ice coats roads making driving dangerous. Too much rain can cause flooding, which has the potential for destroying homes and killing people. Too little rain can result in droughts, which have the potential to cause living things to dehydrate and eventually die. Water can carry pathogens, which cause disease; in developing countries waterborne diseases are commonplace.

WATER'S CHEMICAL COMPOSITION

Water is a molecule comprised of two hydrogen (H) atoms and one oxygen (O) atom. (See Figure 1.1.) The atoms in a molecule of water share electrons, forming strong chemical bonds that hold water molecules together. Water is also both a weak acid and a weak base—chemical properties that allow it to dissolve many substances.

THREE STATES OF WATER

Water exists naturally in three states: a liquid (its most common form), a solid (ice), and a gas (water vapor). It is the only substance on Earth in which all three of its natural states occur within the normal range of climatic conditions, sometimes at the same time. Familiar examples of water in its three natural states are rain, snow or hail, and steam.

Compared with other liquids, water has some unusual properties. For example, most liquids contract (shrink) as they freeze. That is, their molecules move closer together. Water contracts only until it reaches 4° Celsius (C) or 39.2° Fahrenheit (F). Then it expands (its molecules move farther apart from one another) until it reaches its freezing point of 0°C (32°F). This expansion can exert a tremendous force on surrounding objects, enough to crack an unprotected automobile engine, burst a basement water pipe, or even shatter a boulder. Expansion makes ice less dense than water, which is why ice floats. This phenomenon causes

FIGURE 1.1

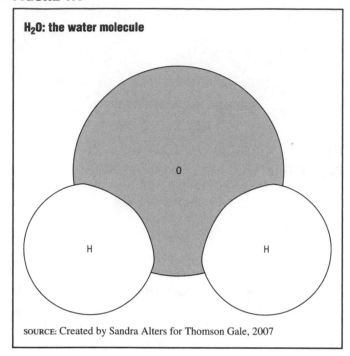

H₂0: the water molecule

ice to form on rivers and lakes from the top down, and aquatic life survives beneath the ice.

When ice is warmed to a temperature higher than 0°C, it melts, becoming liquid. As a liquid, its molecules are more loosely bound together than in the "locked" crystalline lattice of ice, and they can move around each other rather freely. The molecules' ability to slip and slide around gives water and other liquids their fluid properties.

In the gaseous or vapor state, water molecules move rapidly about and have little attraction for each other, creating the diffuse appearance of steam or mist, or the haze of a humid day. (Humidity is the measure of the amount of water vapor in the air.) Evaporation is the general term used to describe the process by which water in its liquid form is changed to its gaseous state. Evaporation can occur under a wide variety of conditions. Examples include water vaporizing off of wet pavement following rainfall, boiling in a pot on the stove and producing steam, or evaporating from clothes hanging on a line to dry.

WATER: THE EARTH MOVER

Scientists believe the earth formed about four billion years ago. Within its primitive atmosphere were the basic elements needed to form water. As the earth cooled from a mass of molten rock, water formed in the atmosphere and then fell to the ground in a rain that lasted for many years, forming the oceans.

The flow of water flattens mountains and cuts canyons deep into the surface of the earth. It hollows out underground caverns and leaves behind attractive forma-

tions. Water creates soil by breaking down rocks and organic material and depositing it elsewhere. Water in the form of ice redesigned the face of the earth as glaciers advanced and receded many thousands of years ago. The slow, relentless processes of water freezing, melting, flowing, and evaporating will likely make the earth's appearance as different a million years from today as it was a million years ago.

HYDROLOGIC CYCLE

The U.S. Geological Survey (USGS) notes in *Where Is Earth's Water Located?* (August 28, 2006, http://ga.water.usgs.gov/edu/earthwherewater.html) that the earth is a vast reservoir, containing an estimated 332.5 million cubic miles of water (a cubic mile of water equals 1.1 trillion gallons). It is all around us: in the atmosphere, on the earth's surface, and in the ground. The relative distribution of the world's water supply is shown in Figure 1.2. About 97% of water on Earth is saltwater in the oceans, and about 3% is freshwater. Of the freshwater, 68.7% comprises ice caps and glaciers, 30.1% is groundwater (water found within the ground), and 0.3% is surface water (water at the surface, such as lakes and rivers). Only 0.9% of water is found in the atmosphere, mainly in the form of invisible water vapor.

Of all this water, what can humans use for their daily water needs? The 97% of water found in the oceans cannot be used unless the salt is removed. The desalination process is quite costly, although recent technologies have made this process economically more attractive. Any community, business, or industry that considers desalination as a method of obtaining freshwater must determine whether the cost of desalination is lower than the cost of other water supply alternatives. Most of the desalination facilities currently in operation around the world are in the Middle East. The article "Largest Desalination Plant Opens in Ashkelon" (*Israel High-Tech and Investment Report*, September 2005) explains that in August 2005 Israel began operation of the world's largest desalination plant. Located along the country's southern Mediterranean coast, it provides a hundred million cubic meters of desalinated water per year, about 15% of the total household water in Israel. In the United States, Florida and California have desalination plants. The California Coastal Commission indicates in *Seawater Desalination and the California Coastal Act* (March 2004, http://www.coastal.ca.gov/energy/14a-3-2004-desalination.pdf) that California's 11 plants provide about 3 million gallons per day and that the state plans on building 21 more facilities, which will provide about 240 million gallons per day.

Most of the water people use everyday, however, comes from rivers, which is only 2% of all surface water, and surface water makes up only 0.3% of all the water on

FIGURE 1.2

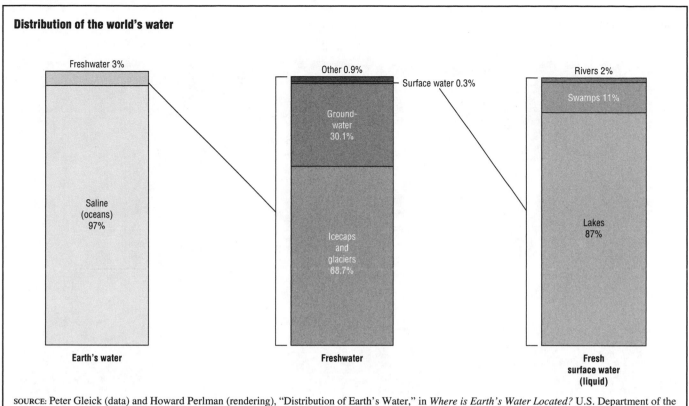

Distribution of the world's water

SOURCE: Peter Gleick (data) and Howard Perlman (rendering), "Distribution of Earth's Water," in *Where is Earth's Water Located?* U.S. Department of the Interior, U.S. Geological Survey, August 28, 2006, http://ga.water.usgs.gov/edu/earthwherewater.html (accessed January 2, 2007)

Earth. (See Figure 1.2.) Most of the freshwater available for use (30.1% of all freshwater) is stored in the ground. The rest of the freshwater on Earth is not available for human use because it is frozen in icecaps and glaciers.

As shown in Figure 1.3, the earth's water is continually being cycled between the ground and the air. This exchange is caused by the heat of the sun and the force of gravity. Water evaporates from the surface of the earth (e.g., from moist ground, the leaves of vegetation, and bodies of water). It then rises into the atmosphere as water vapor. The water vapor condenses to a liquid from its gaseous form and falls as precipitation, in the form of rain, mist, sleet, hail, or snow.

The precipitation, in turn, replenishes the earth's surface and underground waters, which eventually join the ponds, lakes, rivers, and oceans. Evaporation of water from the land and oceans and transpiration from plants (evaporation of water from the leaves), which together are called evapotranspiration, put the water (as vapor) back into the air. In this way water travels from the ground to the atmosphere and back to the ground continuously. The exchange of water between the ground and the air is called the hydrologic cycle or water cycle. (See Figure 1.3.) The term *hydrologic* is derived from two Greek words: *hydro*, which means water, and *loge*, an ancient Greek word meaning "knowledge of."

The hydrologic cycle is a natural, constantly running distillation and pumping system. As a cycle, this flow has no beginning and no end. Within the hydrologic cycle water is neither lost nor gained; it simply changes form as it moves through the cycle. The molecules of water in the world's oceans, lakes, rivers, ponds, streams, and atmosphere today are the same molecules that formed four billion years ago.

Although constantly in motion, water is transferred between phases of the hydrologic cycle at different rates, depending on where it is located. For instance, a water molecule exists as water vapor in the atmosphere an average of eight days, but when it enters the ocean it may remain there for the next twenty-five hundred years.

The hydrologic cycle shapes and sustains life on Earth. It is largely responsible for determining climate and types of vegetation. Because it is an open system, outside actions may affect any phase of the cycle, and they may have both immediate and long-term consequences.

Soil Moisture

Although it represents only a small percentage of Earth's water supply, soil moisture is extremely significant. It supplies water to plants, a vital link in the food chain. Some plants grow directly in water or in marshy ground, but most live on dry land. This is possible because the land is truly dry in just a few places, and often only temporarily.

FIGURE 1.3

The water cycle

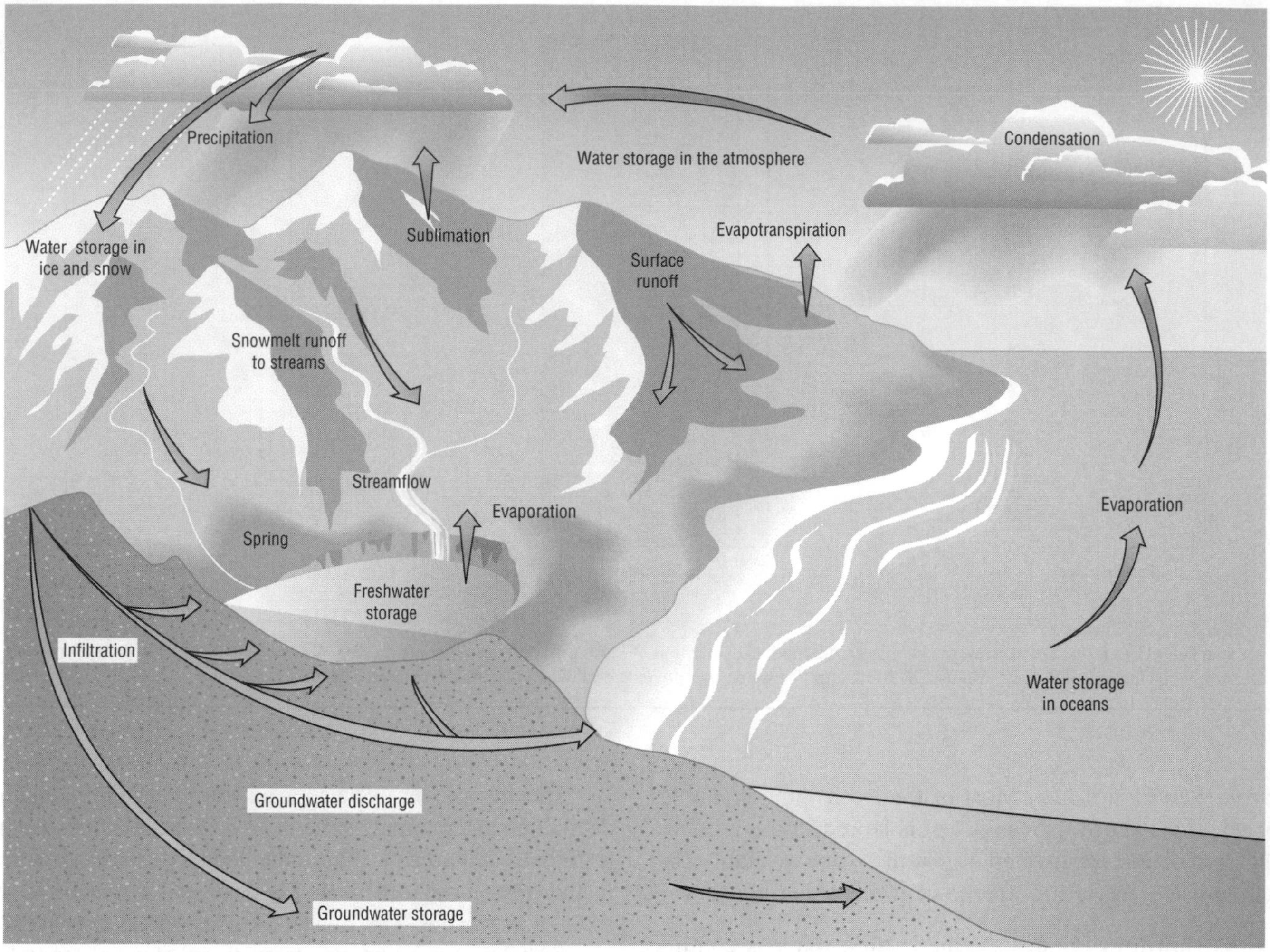

Precipitation

Water storage in the atmosphere

Condensation

Sublimation

Evapotranspiration

Water storage in ice and snow

Surface runoff

Snowmelt runoff to streams

Streamflow

Evaporation

Spring

Evaporation

Freshwater storage

Infiltration

Water storage in oceans

Groundwater discharge

Groundwater storage

SOURCE: John M. Evans (illustrator) and Howard Perlman (rendering), "The Water Cycle," in *What is the Water Cycle?* U.S. Department of the Interior, U.S. Geological Survey, October 6, 2006, http://ga.water.usgs.gov/edu/watercyclehi.html (accessed January 2, 2007)

Dust is generally considered dry, but the dust kicked up by a car on a dry dirt road may contain up to 15% water by weight. Vegetation, however, cannot grow and flourish on that road because soil holds its small percentage of moisture so strongly that plant roots cannot get it out. Other than desert plants, which store water in their own tissues during infrequent wet periods, most plants can grow only where there is extractable water in the soil. Because the earth's vegetation continually withdraws moisture from the ground in large amounts, frequent renewals of soil moisture, either by precipitation or irrigation, are needed.

Atmospheric Moisture

Rain, snow, sleet, and hail are all forms of precipitation. This moisture in the air comes from the evaporation of water from the ground and from bodies of water such as lakes, rivers, and, especially, the oceans. Plants also release moisture into the air through their leaves. This process is called transpiration. The plant moisture is first taken up by the roots from the soil, moves up the plant in the sap, and then emerges from the plant through thousands of tiny holes on the underside of each leaf. According to the USGS, in "The Water Cycle: Evapotranspiration" (August 25, 2005, http://ga.water.usgs.gov/edu/watercycleevapotranspiration .html), an acre of corn transpires three to four thousand gallons of water every day and a large tree may release over one hundred gallons per day.

Transpiration from plants is one of the important sources of water vapor in the air and usually produces more moisture than evaporation from the ground, lakes, and streams. The most important source of water vapor in the air, however, is evaporation from the oceans, especially those parts of the ocean that are located in the warmest parts of the planet. Heat is required to change water from a liquid to a vapor. Thus, the higher the temperature, the faster the water evaporates from the oceans. However, the winds in the upper atmosphere carry this moisture far from the oceans. Someone who lives in the central part of the United States, for example, may receive rain that is composed of water particles evaporated from the ocean near the equator or the Gulf of Mexico.

Ice Caps and Glaciers

About two-thirds of all freshwater in the world is stored as ice. (See Figure 1.2.) According to the article "Antarctic Ice Sheet Losing Mass, Says University of Colorado Study" (*ScienceDaily*, March 2, 2006), most of the world's ice is held by the Antarctic ice cap—about 90% of all existing ice. However, this ice sheet has lost significant mass in recent years due to changing climate conditions.

Ice is also held in glaciers. A glacier is any large mass of snow or ice that persists on land for many years and moves under its own weight. Glaciers are formed in locations where, over a number of years, more snow falls than melts. As this snow accumulates, it compresses and changes into dense, solid ice—a glacier.

Robert M. Krimmel notes in *Glaciers of the Coterminous United States* (March 7, 2002, http://pubs.usgs.gov/pp/p1386j/us/westus-lores.pdf) that even though most people associate glaciers with remote, frozen regions such as Antarctica, there are more than sixteen hundred glaciers in the lower forty-eight U.S. states, most of them quite small. They cover about 227 square miles in parts of California, Colorado, Idaho, Montana, Nevada, Oregon, Utah, Washington, and Wyoming. Alaska has uncounted numbers of glaciers that cover many thousands of square miles.

Permafrost

Permafrost is permanently frozen ground, which comprises approximately one-fifth of Earth's entire land surface. It exists in Antarctica but is more extensive in the Northern Hemisphere. In the land surrounding the Arctic Ocean, its maximum thickness has been measured in thousands of feet—about 5,000 feet (1,524 meters) in Siberia and 2,000 feet (610 meters) in Alaska. Even though the surface permafrost (called the active layer, see Figure 1.4) thaws quickly in the summer and refreezes in winter, it would take thousands of years of thawing conditions to melt the thick, frozen layer beneath. The deepest part of permafrost was formed during the Great Ice Age, which occurred about three million years ago. Talik, which is shown in Figure 1.4, are patches of ground within permafrost that never freeze because of a local irregularity in conditions.

In North America the discovery of gold in Alaska and the Yukon in the early 1900s sparked an increased

FIGURE 1.4

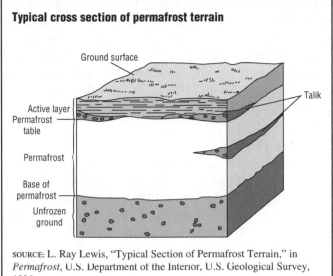

Typical cross section of permafrost terrain

SOURCE: L. Ray Lewis, "Typical Section of Permafrost Terrain," in *Permafrost*, U.S. Department of the Interior, U.S. Geological Survey, 1986

interest in the nature of the vast areas of permafrost. After World War II (1939–45), increasing numbers of nonnative people migrated to areas of frozen ground, and the construction of roads, railroads, and buildings and the clearing of land led to the disruption and thawing of previously undisturbed permafrost. This caused unstable ground, landslides, mudflows, and, consequently, dangerous living conditions. In addition, Larry C. Smith et al. report in "Disappearing Arctic Lakes" (*Science*, June 3, 2005) that warming in the Arctic, which has sped up since the 1980s, is causing the loss of permafrost in Arctic regions.

Snowmelt

Except for the disruption of day-to-day life caused by winter snowstorms in certain areas of the United States, most Americans are largely unaware of the importance of snow. Unlike many other countries, the United States is economically dependent on snow. Almost all the water in the arid West that can be tapped on a large-volume basis comes directly from spring snowmelt. The amount of water in a given year's snowpack varies greatly from one year to another. The snowpack volume is of crucial importance to regional economics. Too much snow can cause flooding and extensive damage to crops, livestock, businesses, and homes. Too little can mean shortages in water for drinking, irrigation, and hydroelectric power, affecting their availability and cost.

The importance of snow is highlighted by Mark H. Hunter in "San Luis Valley Still in the Grip of Record Drought. Snowpack Goes into Ground, Not Streams" (*Denver Post*, May 2, 2003). Hunter describes drought conditions in northern Colorado, caused in part by a significant decrease in snowpack in the upper Rio Grande basin. According to Hunter, in 2003 snowpack in the upper

Rio Grande basin was only 71% of average, with a snow-water content of thirteen inches. Drought conditions had affected the region for three consecutive years, diminishing the San Luis Valley's underground aquifer. The aquifer sustains thousands of acres of natural wetlands and half a million acres of farm and ranch land. The area's snowpack improved somewhat in 2004 and 2005, but the effects of the drought are still being felt.

FRESHWATER

Despite the enormous amount of water that surrounds us, only about 3% of it is freshwater and, therefore, suitable for use by land-based animals, plants, and humans. (See Figure 1.2.) The availability of freshwater depends on many factors: climate, location, rainfall, and local activity. The USGS reports in *Where Is Earth's Water Located?* that the world's freshwater lakes contain 21,830 cubic miles of water, and the world's rivers contain 509 cubic miles of freshwater. Together, lakes and rivers make up about 0.0072% of the total water on Earth. Groundwater (subsurface water) totals about 5.6 million cubic miles or 1.7% of the total water supply.

VARIATIONS IN PRECIPITATION

The amount of precipitation that falls around the world can range from less than one-tenth of one inch per year in the deserts to hundreds of inches per year in the tropics. In "Global Measured Extremes of Temperature and Precipitation" (August 9, 2004, http://www.ncdc.noaa.gov/oa/climate/globalextremes.html#highpre), the National Climatic Data Center reports that the lowest average annual precipitation occurs in Arica, Chile, with 0.03 inches. The world's wettest spot is Lloro, Colombia, with an average annual rainfall of 523.6 inches.

Variations in precipitation occur not only in various regions of the globe but also seasonally and annually. For example, southern Florida has a rainy season (May to October) followed by a dry season (November to April). Most of the forty-five to sixty inches of annual rain that falls (under normal conditions) in this area occurs in the rainy season. In exceptionally dry years, droughts occur because the area receives little or no precipitation in the rainy season; in exceptionally wet years, flooding may occur.

Natural phenomena known as El Niño and La Niña influence weather and precipitation. El Niño is a naturally occurring disruption of the ocean-atmosphere system in the tropical Pacific Ocean, which has important consequences for weather around the globe. It is characterized by an unusually warm current of water that appears every three to five years in the eastern Pacific Ocean. Unusually warm sea surface temperature results in a decline in primary productivity (microscopic plants and animals) that in turn brings sharp declines in commercial fisheries and bird populations that are also dependent on fish. Unusual weather conditions occur around the globe as jet streams, storm tracks, and monsoons are shifted. Some other consequences are increased rainfall across the southern United States and Peru that has caused destructive flooding in the past, and drought in Australia and Indonesia. El Niño brings warmer than normal temperatures to the north-central states and cooler than normal temperatures to the southeastern and southwestern United States.

La Niña global climate impacts tend to be the opposite of El Niño because La Niña is characterized by unusually cold ocean temperatures in the equatorial Pacific. In the United States winter temperatures are warmer than normal in the Southeast and cooler than normal in the Northwest. La Niña events occur after some, but not all, El Niño events. Generally, La Niña occurs half as frequently as El Niño.

HUMAN INFLUENCES ON WATER

As populations continually modify the environment to suit their needs and desires, the natural processes, including the hydrologic cycle, are significantly disrupted. People are finding out that the earth, even with its remarkable recuperative powers, has limits beyond which it cannot sustain a livable environment.

There are two ways by which humanity can change the basic quality and natural distribution of water: by introducing materials and organisms into a body of water (including the atmosphere)—commonly known as pollution—and by intervening in any phase of the hydrologic cycle in such a way that the cycle is altered. Dams, irrigation, and hydroelectric plants are examples of alterations.

Water Pollution

For centuries, the world's lakes, rivers, and oceans have been dumping sites for many of the undesirable byproducts of civilization. People have dumped indiscriminately, believing that bodies of water had an inexhaustible capacity to disperse and neutralize any amount of waste. What was not dissolved or dispersed settled to the bottom, where it could not be seen.

Dumping waste into the oceans and waterways led to few apparent problems as long as waste products were few and consisted mainly of naturally occurring materials. However, as the world's population grew and technology began introducing huge numbers of new products and processes, this natural disposal system began breaking down under an overload of natural and synthetic contaminants. Fish and marine animals died; dead zones, where no life could survive, developed in harbors and oceans; drinking water became contaminated; and beaches became littered with garbage.

TABLE 1.1

Sources of water pollution

Category	Examples
Industrial	Pulp and paper mills, chemical manufacturers, steel plants, metal process and product manufacturers, textile manufacturers, food processing plants
Municipal	Publicly owned sewage treatment plants that may receive indirect discharges from industrial facilities or businesses
Combined sewer overflows	Single facilities that treat both storm water and sanitary sewage, which may become overloaded during storm events and discharge untreated wastes into surface waters
Storm sewers/ urban runoff	Runoff from impervious surfaces including streets, parking lots, buildings, and other paved areas
Agricultural	Crop production, pastures, rangeland, feedlots, animal operations
Silvicultural	Forest management, tree harvesting, logging road construction
Construction	Land development, road construction
Resource extraction	Mining, petroleum drilling, runoff from mine tailing sites
Land disposal	Leachate or discharge from septic tanks, landfills, and hazardous waste sites
Hydrologic modification	Channelization, dredging, dam construction, flow regulation
Habitat modification	Removal of riparian vegetation, streambank modification, drainage/filling of wetlands

SOURCE: "Table 1-1. Pollution Source Categories Used in This Report," in *National Water Quality Inventory: 1998 Report to Congress*, U.S. Environmental Protection Agency, Office of Water, June 2000, http://www .epa.gov/305b/98report/chap1.pdf (accessed January 4, 2007)

Water that has been physically or chemically changed and that adversely affects the health of humans and other organisms is said to be polluted. There are many sources and types of water pollution. Every day, industrial byproducts and household wastes such as toxic chemicals, metals, plastics, medical refuse, radioactive waste, and sludge (the solid material left after water is extracted from raw sewage) are deposited into the nation's rivers, lakes, harbors, and oceans. Septic tanks, landfills, and mining

operations often produce hazardous substances that seep into the soil and then into underground aquifers (areas within subsurface rock where water is stored). Table 1.1 lists the most common sources of water pollution. (The term *riparian*, which is used in Table 1.1, refers to the banks of a body of water, such as a riverbank.) Figure 1.5 shows how bacteria, viruses, and other pathogens can be introduced into water.

Water pollution is acknowledged by both scientists and the WHO as a global problem and one that results in the serious health issue of waterborne disease. Concern over water pollution helped launch the environmental movement of the 1970s. The 1972 Federal Water Pollution Control Act, commonly known as the Clean Water Act, was the first major piece of environmental legislation enacted by Congress. Since then, many laws and regulations designed to protect, preserve, and clean up the national waters have been passed. Although substantial progress has been made, many problems remain to be solved.

Point and Nonpoint Sources of Pollution

There are two types of water pollution sources: point and nonpoint sources. Point sources are specific sites, such as sewage treatment plants, factories, and ships, which discharge pollutants into bodies of water at single points via pipes, sewers, or ditches. Nonpoint sources are not specific sites; pollutants enter bodies of water over large areas rather than at single points. Nonpoint sources of water pollution include agricultural runoff, mining activities, and soil erosion.

The Water Pollution Control Act and its amendments, such as the Clean Water Act, established the National Pollutant Discharge Elimination System, which controls

FIGURE 1.5

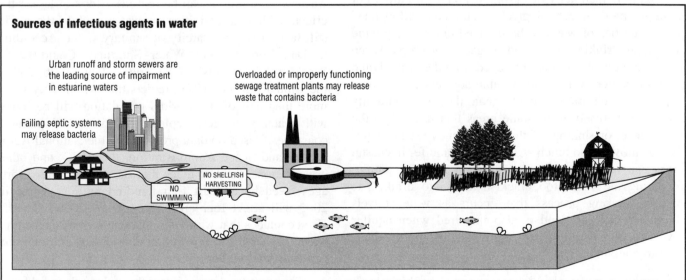

Sources of infectious agents in water

Urban runoff and storm sewers are the leading source of impairment in estuarine waters

Overloaded or improperly functioning sewage treatment plants may release waste that contains bacteria

Failing septic systems may release bacteria

NO SHELLFISH HARVESTING

NO SWIMMING

SOURCE: "Fig. 5-5. Sources of Bacteria," in *National Water Quality Inventory: 1998 Report to Congress*, U.S. Environmental Protection Agency, Office of Water, June 2000, http://www.epa.gov/305b/98report/chap5.pdf (accessed January 4, 2007)

water pollution by regulating point sources. This system uses water quality standards and discharge permits as a means of regulation. Water quality standards establish the upper limit for the amount of a pollutant that will not cause an adverse effect on humans or other species. Cities, companies, and other entities that want to discharge into water apply for permission and if approved receive a permit. The permit specifies the amount and type of pollutants that may be discharged and not cause a violation of the water quality standard. Dischargers are required to monitor what they release and report the results. When limits are exceeded, fines and other penalties are imposed, including requirements for additional treatment and cleanup.

Nonpoint sources of pollution are harder to control. Agriculture results in a great deal of nonpoint source pollution. For example, when water runs off fertilized land, it carries fertilizer with it into bodies of water. Fertilizer promotes the growth of plants and algae in the water. A high concentration of plant and algal growth results in the water becoming cloudy with their growth and, therefore, light cannot easily penetrate. When the reduced levels of light and the concentration of nutrients can no longer support high plant and algal growth, these organisms die, fall to the bottom, and decay. As bacteria feed on the dead plant material, they use oxygen, which results in less dissolved oxygen for fish. Fish that need higher levels of oxygen die out, whereas other species that can survive in low oxygen levels multiply. The dead fish decay as well—a process that lowers the dissolved oxygen even more. At this point, the water is said to be eutrophic and may become slimy and smelly.

Human Activities That Contribute to Water Pollution and Degrade Water Quality

Many human activities promote water pollution. For example, modern technological developments allow massive quantities of water to be pumped out of the ground for use as drinking water and irrigation of crops. When large amounts of water are removed from the ground (and from the water cycle), underground aquifers can become depleted much more quickly than they can naturally replenish themselves. In some areas this has led to the subsidence, or sinking, of the ground above major aquifers. Removing too much water from an aquifer in coastal areas can result in saltwater intrusion into the aquifer, rendering the water too brackish (salty) to drink. The natural filtering process that occurs as water travels through rocks and sand is also impaired when aquifer levels become depleted, leaving the aquifer more vulnerable to contamination.

Building dams also interferes with the hydrologic cycle and may promote water pollution. The huge dams built in the United States just before and after World War

II have substantially changed the natural flow of rivers. By reducing the amount of water available downstream and slowing stream flow, a dam not only affects a river but also the river's entire ecological system. For example, wetlands have the ability to clean water by trapping and filtering pollutants. This water-cleansing process can be stopped or reduced if dams cause wetlands to dry up.

Deforestation and overgrazing worldwide have destroyed thousands of acres of vegetation that play a vital role in controlling erosion. Erosion is the process by which a material is worn away by a stream of water or air, usually because of the abrasive particles in the water or air. Erosion results in soil runoff into rivers and streams, causing turbidity (cloudiness or discoloration), siltation (depositing of soil on the bottoms of rivers and streams), and disruption of stream flow. Removal of vegetation on the land also reduces the amount of water released into the atmosphere by transpiration. In some areas less water in the atmosphere can mean less rainfall, causing fertile regions to become deserts.

Along with agricultural activities, industrial, urban, and residential development can also lead to soil runoff. As Figure 1.6 shows, timber harvesting leads to compacted soil, less ground cover, and disturbed ground. Some agricultural and industrial practices also lead to these results, along with fewer riparian areas. Building roads results in compacted soil and fewer floodplains and wetlands. Urbanization (the building of cities with paved surfaces covering the land) leads to all these results. Under these conditions storms often lead to increased turbidity, flooding, runoff, and erosion.

WORLDWIDE WATER CRISIS

Despite conservation (the careful use and protection of water resources) and reclamation (the treatment of wastewater so that it can be reused) efforts to lessen the effects of human activities on water quality, the world still faces a severe scarcity of sanitary water. According to Dan Vergano, in "Water Shortages Could Leave World in Dire Straits" (*USA Today*, January 26, 2003), the United Nations (UN) predicts that within fifty years more than half of the world's population will be living with water shortages, depleted fisheries, and polluted coastlines. Based on data provided by the National Aeronautics and Space Administration, the WHO, and other organizations, the UN notes that the severe water shortages faced by at least four hundred million people in 2003 will affect four billion people by 2050. The UN also concludes that waste and inadequate management of water, especially in poverty-stricken areas, are the main causes of the problems.

The scarcity of clean water throughout the world also has significant implications for public health. The UN Educational, Scientific, and Cultural Organization (2006,

FIGURE 1.6

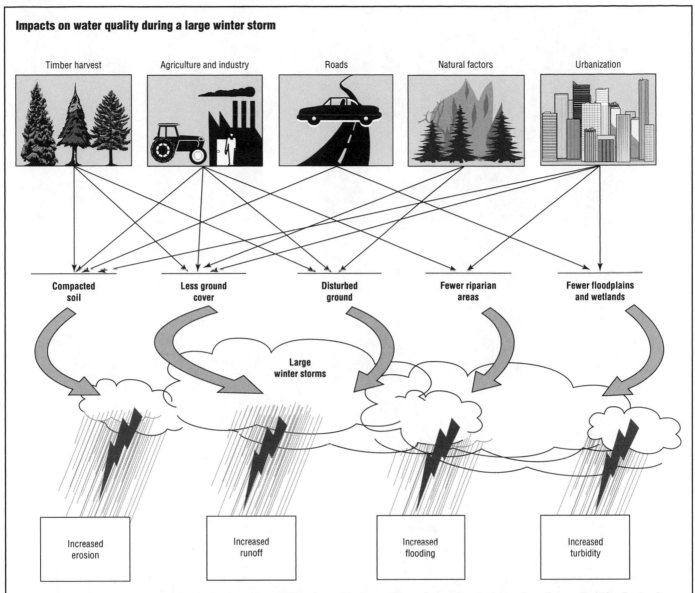

Impacts on water quality during a large winter storm

Timber harvest   Agriculture and industry   Roads   Natural factors   Urbanization

Compacted soil   Less ground cover   Disturbed ground   Fewer riparian areas   Fewer floodplains and wetlands

Large winter storms

Increased erosion   Increased runoff   Increased flooding   Increased turbidity

SOURCE: "Figure 2. Impacts on Water Quality During a Large Winter Storm," in *Oregon Watersheds: Many Activities Contribute to Turbidity During Large Storms*, U.S. General Accounting Office, July 1998, http://www.gao.gov/archive/1998/rc98220.pdf (accessed January 4, 2007)

http://www.unesco.org/water/wwap/facts_figures/basic_needs.shtml) finds that water contaminated with bacteria, parasites, and other microbes causes about 6,000 deaths every day, including 1.3 million children under the age of five.

Water shortages also affect international politics. In *Global Water Futures* (2005, http://www.sandia.gov/water/docs/Global_Water_Futures.pdf), the Center for Strategic and International Studies estimates that more than 260 of the world's river basins are shared by at least two countries. Conflicts arise when one country tries to dam up, siphon off, or pollute the shared water. According to Meredith A. Giordano and Aaron T. Wolf, in *Atlas of International Freshwater Agreements* (November 2003, http://www.transboundarywaters.orst.edu/publications/

atlas/), since 1820 there have been more than four hundred agreements related to water as a limited and consumable resource. They note that while cooperation about water resources between and among countries during the past fifty years has outnumbered conflicts by more than two to one, problems still occur. For example, between 1948 and 2002 Giordano and Wolf count thirty-seven incidents of violent conflict over water, thirty of which were between Israel and one or another of its neighbors. Peter H. Gleick, in "Water Conflict Chronology" (October 12, 2006, http://worldwater.org/conflictchronology .pdf), identifies several more instances between 2003 and 2006, including disputes in Colombia, Ethiopia, India, Iraq, Israel, Kenya, Lebanon, Mexico, Pakistan, Somalia, Sri Lanka, Sudan, and Yemen.

CHAPTER 2
WATER USE

Water is used in every society. Individuals use water for drinking, cooking, cleaning, and recreation. Industry uses it to make chemicals, manufacture goods, and clean factories and equipment. Cities use water to fight fires, clean streets, and fill public swimming pools and fountains. Farmers give water to their livestock, clean their barns, and irrigate their crops. Hydroelectric power stations use water to drive generators, and thermonuclear power stations use it for cooling. No plant or animal can survive without water. Water is vital to life, yet, as Chapter 1 describes, it is a finite (limited) resource. There is no more water today than was on Earth four billion years ago.

FRESHWATER AVAILABILITY

Most human and land-based animal and plant activities that use water require freshwater. In the vast majority of cases, saline or saltwater cannot be used without treating it to remove the salt. The U.S. Geological Survey (USGS) notes in *Where Is Earth's Water Located?* (August 28, 2006, http://ga.water.usgs.gov/edu/earthwherewater.html) that the world's total water supply is about 332.5 million cubic miles, but freshwater comprises only about 3% of this water. (See Figure 1.2 in Chapter 1.) If this water were distributed equally over the planet relative to population density and animal and plant needs, it would be more than enough to sustain all life. This, however, is not the case.

Freshwater supplies vary not only from region to region on Earth but also from year to year within regions. Within the continental United States some parts of the country do not have adequate supplies at the same time that other areas may be experiencing floods. For example, the National Climatic Data Center, in *Climate of 2006—In Historical Perspective: Annual Report* (January 9, 2007, http://www.ncdc.noaa.gov/oa/climate/research/2006/ann/ann06.html), reports that as rains flooded western Washington State in November 2006, parts of Arizona and Minnesota were experiencing a severe drought.

The first human settlements were based on the availability of water. Where water was plentiful, large numbers of people flourished; where water was scarce, small groups eked out a living. Villages and cities thrived in areas of constant water supply. In more arid regions nomads wandered in search of water. Great nations grew up along the Nile River in Egypt, the Tigris and the Euphrates rivers in western Asia, the Indus River in India, and the Yellow River in China.

Modern societies, which have more control over the water supply than did ancient societies, have developed technologies that bring water to arid regions and divert water from areas likely to flood. Modern, elaborate irrigation systems have made it possible for cities to exist in places where two centuries ago only the hardiest plants and animals could survive. For example, without these water systems Los Angeles would be a semiarid desert.

HOW WATER IS SUPPLIED

Freshwater that is potable (safe to drink) is the most crucial resource for the maintenance of human societies. Freshwater, however, is limited in total supply, unevenly distributed, and often of unacceptable quality, particularly in areas where the supply is limited.

Most people in the United States obtain water through water utility companies, also called water purification and distribution plants. Utility companies are those that serve the public, such as an electric company, sewage treatment plant, or water purification and distribution plant. Utility companies may be owned by cities, towns, or private entities.

Water utility companies withdraw water from either surface or groundwater sources to supply their customers. The customers pay the utility companies for the water they use. Water may also be self-supplied, that is, withdrawn directly from wells, lakes, or rivers by those users

FIGURE 2.1

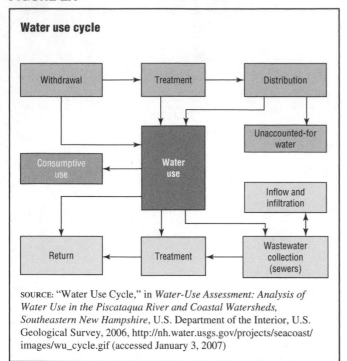

Water use cycle

SOURCE: "Water Use Cycle," in *Water-Use Assessment: Analysis of Water Use in the Piscataqua River and Coastal Watersheds, Southeastern New Hampshire*, U.S. Department of the Interior, U.S. Geological Survey, 2006, http://nh.water.usgs.gov/projects/seacoast/images/wu_cycle.gif (accessed January 3, 2007)

who have the equipment, technology, and water rights necessary to withdraw and process water for their individual and family use.

As described in Chapter 1, water cycles naturally within the environment. (See Figure 1.3 in Chapter 1.) Water withdrawn from surface or groundwater sources by humans eventually returns to the water cycle, but sometimes it is returned in a condition different from which it was withdrawn. The water's condition can greatly affect the ability to reuse the water. For example, water used for irrigation may end up as agricultural runoff containing pesticides and fertilizers, making it unfit for other uses such as drinking. Water used to flush a toilet mixes with body wastes, making it unfit for other uses as well.

Figure 2.1 shows a flowchart of water use activities. The phrase *water use* means human activities that use and transfer both surface and groundwater. Water use begins when water is withdrawn from groundwater or surface water. It may be withdrawn and used directly, such as when a person has a well serving the home, or it may be withdrawn, treated, and distributed by a water utility company. After use, water that is not consumed is called wastewater, and it is returned to a river or to a wastewater treatment plant where, after treatment, it is discharged into a river. Consumed water is that which evaporates during water use or is used in a way that makes it unavailable for other uses, such as water incorporated into a product being manufactured. Unaccounted-for water includes public water use (firefighting, street wash-ing, water treatment plant back flushing of filters, and municipal parks and swimming pools), leakage (conveyance loss), and meter errors.

TYPES OF WATER USE IN THE UNITED STATES

Water use in the United States is monitored and reported by the USGS, which classifies water use as in-stream or off-stream. In-stream use means that water is used at its source—usually a river or stream—and the vast majority of that water is returned immediately to the source. Little or no water is consumed. Examples are a hydroelectric plant where water flows through turbines, which drive generators, and old mills where flowing water turns a wheel, which moves grinding stones. In the hydroelectric plant, the flowing water immediately returns to the river from which it came. In the old mill, the water never leaves the river. Off-stream use means that the water is withdrawn from a surface or groundwater source and conveyed to the place where it is used.

Off-Stream Use

Table 2.1 shows total off-stream water use by source, type, and state for 2000. Nationwide, off-stream water use was 408 billion gallons per day (Bgal/d). Susan S. Hutson et al., in *Estimated Use of Water in the United States in 2000* (2004, http://pubs.usgs.gov/circ/2004/circ1268/pdf/circular1268.pdf), the most recent info available, note that this was slightly more than the 402 Bgal/d in 1995. Of the water involved in off-stream use, 345 Bgal/d (84%) was freshwater in 2000.

At 51.2 Bgal/d, California used more water than any other state by a large margin. (See Table 2.1.) Texas was the second-largest water user at 29.6 Bgal/d. Figure 2.2 shows a map of total off-stream water use by state for 2000. This presentation of the data makes it easier to compare water use among states. It is easy to see that California, Texas, and Florida are the largest water-users in the nation.

In many cases off-stream use results in substantial consumptive use of water. Consumptive use is of two types: quantitative or qualitative. Quantitative consumption means that part of the water withdrawn has evaporated, been transpired (given off) by plants, been incorporated into products or crops, been consumed by humans or livestock, or otherwise removed from the immediate water environment so that the quantity returned to the source is substantially less than the quantity of water withdrawn. An example of consumptive use is spray irrigation. According to the USGS, in "Irrigation Techniques" (August 30, 2005, http://ga.water.usgs.gov/edu/irmethods.html), about 60% of the water used to irrigate crops is returned to the source. The other 40% of the water evaporates, is incorporated into plant structure, or is transpired by plants.

TABLE 2.1

Total water use by source and state, 2000

[Figures may not sum to totals because of independent rounding]

State	Population (in thousands)	Groundwater Fresh	Groundwater Saline	Groundwater Total	Surface water Fresh	Surface water Saline	Surface water Total	Total Fresh	Total Saline	Total Total	Acre-feet Fresh	Acre-feet Saline	Acre-feet Total
		Withdrawals (in million gallons per day)									Withdrawals (in thousand acre-feet per year)		
Alabama	4,450	440	0	440	9,550	0	9,550	9,990	0	9,990	11,200	0	11,200
Alaska	627	50.2	90.4	141	111	53.4	164	161	144	305	181	161	342
Arizona	5,130	3,420	8.17	3,430	3,300	0	3,300	6,720	8.17	6,730	7,530	9.16	7,540
Arkansas	2,670	6,920	.08	6,920	3,950	0	3,950	10,900	.08	10,900	12,200	.09	12,200
California	33,900	15,200	152	15,400	23,200	12,600	35,800	38,400	12,800	51,200	43,100	14,300	57,400
Colorado	4,300	2,320	0	2,320	10,300	0	10,300	12,600	0	12,600	14,200	0	14,200
Connecticut	3,410	143	0	143	565	3,440	4,010	708	3,440	4,150	794	3,860	4,650
Delaware	784	115	0	115	466	741	1,210	582	741	1,320	652	831	1,480
District of Columbia	572	0	0	0	9.87	0	9.87	9.87	0	9.87	11.1	0	11.1
Florida	16,000	5,020	0	5,020	3,110	12,000	15,100	8,140	12,000	20,100	9,120	13,400	22,500
Georgia	8,190	1,450	0	1,450	4,960	91.7	5,060	6,410	91.7	6,500	7,190	103	7,290
Hawaii	1,210	433	.85	434	208	0	208	640	.85	641	718	.95	719
Idaho	1,290	4,140	0	4,140	15,300	0	15,300	19,500	0	19,500	21,800	0	21,800
Illinois	12,400	813	0	813	12,900	0	12,900	13,700	0	13,700	15,400	0	15,400
Indiana	6,080	656	0	656	9,460	0	9,460	10,100	0	10,100	11,300	0	11,300
Iowa	2,930	679	0	679	2,680	0	2,680	3,360	0	3,360	3,770	0	3,770
Kansas	2,690	3,790	0	3,790	2,820	0	2,820	6,610	0	6,610	7,410	0	7,410
Kentucky	4,040	189	0	189	3,970	0	3,970	4,160	0	4,160	4,660	0	4,660
Louisiana	4,470	1,630	0	1,630	8,730	0	8,730	10,400	0	10,400	11,600	0	11,600
Maine	1,270	80.8	0	80.8	423	295	718	504	295	799	565	330	895
Maryland	5,300	225	0	225	1,200	6,490	7,690	1,430	6,490	7,910	1,600	7,270	8,870
Massachusetts	6,350	269	0	269	783	3,610	4,390	1,050	3,610	4,660	1,180	4,050	5,220
Michigan	9,940	734	0	734	9,260	0	9,260	10,000	0	10,000	11,200	0	11,200
Minnesota	4,920	720	0	720	3,150	0	3,150	3,870	0	3,870	4,340	0	4,340
Mississippi	2,840	2,180	0	2,180	632	148	781	2,810	148	2,960	3,150	166	3,320
Missouri	5,600	1,780	0	1,780	6,450	0	6,450	8,230	0	8,230	9,220	0	9,220
Montana	902	188	0	188	8,100	0	8,100	8,290	0	8,290	9,300	0	9,300
Nebraska	1,710	7,860	4.55	7,860	4,390	0	4,390	12,200	4.55	12,300	13,700	5.10	13,700
Nevada	2,000	757	0	757	2,050	0	2,050	2,810	0	2,810	3,140	0	3,140
New Hampshire	1,240	85.2	0	85.2	362	761	1,120	447	761	1,210	501	854	1,350
New Jersey	8,410	584	0	584	1,590	3,390	4,980	2,170	3,390	5,560	2,430	3,800	6,230
New Mexico	1,820	1,540	0	1,540	1,710	0	1,710	3,260	0	3,260	3,650	0	3,650
New York	19,000	893	0	893	6,190	5,010	11,200	7,080	5,010	12,100	7,940	5,610	13,600
North Carolina	8,050	580	0	580	9,150	1,620	10,800	9,730	1,620	11,400	10,900	1,810	12,700
North Dakota	642	123	0	123	1,020	0	1,020	1,140	0	1,140	1,280	0	1,280
Ohio	11,400	878	0	878	10,300	0	10,300	11,100	0	11,100	12,500	0	12,500
Oklahoma	3,450	771	256	1,030	990	0	990	1,760	256	2,020	1,970	287	2,260
Oregon	3,420	993	0	993	5,940	0	5,940	6,930	0	6,930	7,770	0	7,770
Pennsylvania	12,300	666	0	666	9,290	0	9,290	9,950	0	9,950	11,200	0	11,200
Rhode Island	1,050	28.6	0	28.6	110	290	400	138	290	429	155	326	481
South Carolina	4,010	330	0	330	6,840	0	6,840	7,170	0	7,170	8,040	0	8,040
South Dakota	755	222	0	222	306	0	306	528	0	528	592	0	592
Tennessee	5,690	417	0	417	10,400	0	10,400	10,800	0	10,800	12,100	0	12,100
Texas	20,900	8,470	504	8,970	16,300	4,350	20,700	24,800	4,850	29,600	27,800	5,440	33,200
Utah	2,230	1,020	26.5	1,050	3,740	177	3,920	4,760	203	4,970	5,340	228	5,570
Vermont	609	43.2	0	43.2	404	0	404	447	0	447	501	0	501
Virginia	7,080	314	0	314	4,880	3,640	8,520	5,200	3,640	8,830	5,830	4,080	9,900
Washington	5,890	1,470	0	1,470	3,800	39.9	3,840	5,270	39.9	5,310	5,910	44.7	5,960
West Virginia	1,810	90.9	0	90.9	5,060	0	5,060	5,150	0	5,150	5,770	0	5,770
Wisconsin	5,360	813	0	813	6,780	0	6,780	7,590	0	7,590	8,510	0	8,510
Wyoming	494	541	222	763	4,400	0	4,400	4,940	222	5,170	5,540	248	5,790
Puerto Rico	3,810	137	0	137	483	2,190	2,670	620	2,190	2,810	695	2,460	3,150
U.S. Virgin Islands	109	1.03	0	1.03	10.6	136	147	11.6	136	148	13.0	153	166
Total	**285,000**	**83,300**	**1,260**	**84,500**	**262,000**	**61,000**	**323,000**	**345,000**	**62,300**	**408,000**	**387,000**	**69,800**	**457,000**

SOURCE: Susan S. Hutson et al., "Table 1. Total Water Withdrawals by Source and State, 2000," in *Estimated Use of Water in the United States in 2000*, U.S. Department of the Interior, U.S. Geological Survey, 2004, http://pubs.usgs.gov/circ/2004//circ1268/pdf/circular1268.pdf (accessed January 2, 2007) and revision data February 7, 2005, http://pubs.usgs.gov/circ/2004//circ1268/control/revisions.html (accessed January 2, 2007)

Qualitative consumption occurs when the quality of the water is substantially altered so that it is no longer acceptable by downstream users, but the quantity remains substantially unchanged. An example would be discharge of industrial wastewater into a body of water that renders the water unfit for drinking. Many

FIGURE 2.2

Total water use by state, 2000

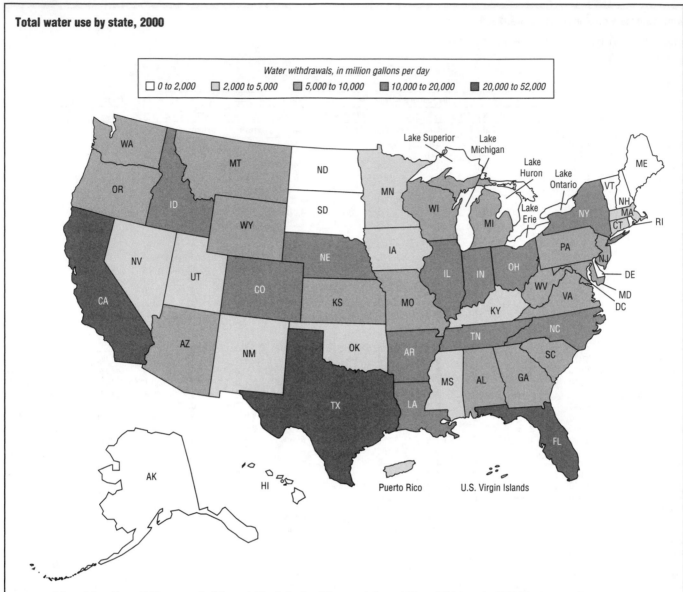

SOURCE: Adapted from Susan S. Hutson et al., "Figure 2. Total, Surface-Water, and Ground-Water Withdrawals, 2000," in *Estimated Use of Water in the United States in 2000*, U.S. Department of the Interior, U.S. Geological Survey, 2004, http://pubs.usgs.gov/circ/2004//circ1268/pdf/circular1268.pdf (accessed January 2, 2007) and revision data February 7, 2005, http://pubs.usgs.gov/circ/2004//circ1268/control/revisions.html (accessed January 2, 2007)

water withdrawals result in both quantitative and qualitative consumption.

Off-stream use is further divided into eight categories:

- Public supply
- Domestic
- Irrigation
- Livestock
- Aquaculture
- Industrial
- Mining
- Thermoelectric power

Table 2.2 shows the total amount of water withdrawn for each category by state. The two activities that use the most freshwater are irrigation (137 Bgal/d nationally in 2000) and thermoelectric power (136 Bgal/d nationally in 2000). Thermoelectric power is the generation of electricity by means of steam-driven turbine generators. No other category of use comes close to using the amount of freshwater that is used for irrigation and thermoelectric power.

PUBLIC-SUPPLY WATER USE. Public-supply water use is water withdrawn by public and private water suppliers (utility companies) and delivered for domestic, commercial, industrial, and thermoelectric power uses. It may be used for public services such as filling public pools,

TABLE 2.2

Total water use by state and usage category, 2000

[Figures may not sum to totals because of independent rounding. All values are in million gallons per day.]

State	Public supply Fresh	Domestic Fresh	Irrigation Fresh	Live-stock Fresh	Aqua-culture Fresh	Industrial Fresh	Industrial Saline	Mining Fresh	Mining Saline	Thermoelectric power Fresh	Thermoelectric power Saline	Total Fresh	Total Saline	Total
Alabama	834	78.9	43.1	—	10.4	833	0	—	—	8,190	0	9,990	0	9,990
Alaska	80.0	11.2	1.01	—	—	8.12	3.86	27.4	140	33.6	0	161	144	305
Arizona	1,080	28.9	5,400	—	—	19.8	0	85.7	8.17	100	0	6,720	8.17	6,730
Arkansas	421	28.5	7,910	—	198	134	.08	2.78	0	2,180	0	10,900	.08	10,900
California	6,120	286	30,500	409	537	188	13.6	23.7	153	352	12,600	38,400	12,800	51,200
Colorado	899	66.8	11,400	—	—	120	0	—	—	138	0	12,600	0	12,600
Connecticut	424	56.2	30.4	—	—	10.7	0	—	—	187	3,440	708	3,440	4,150
Delaware	94.9	13.3	43.5	3.92	.07	59.4	3.25	—	—	366	738	582	741	1,320
District of Columbia	0	0	.18	—	—	0	0	—	—	9.69	0	9.87	0	9.87
Florida	2,440	199	4,290	32.5	8.02	291	1.18	217	0	658	12,000	8,140	12,000	20,100
Georgia	1,250	110	1,140	19.4	15.4	622	30	9.80	0	3,250	61.7	6,410	91.7	6,500
Hawaii	250	12.0	364	—	—	14.5	.85	—	—	0	0	640	.85	641
Idaho	244	85.2	17,100	34.9	1,970	55.5	0	—	—	0	0	19,500	0	19,500
Illinois	1,760	135	154	37.6	—	391	0	—	—	11,300	0	13,700	0	13,700
Indiana	670	122	101	41.9	—	2,400	0	82.5	0	6,700	0	10,100	0	10,100
Iowa	383	33.2	21.5	109	—	237	0	32.8	0	2,540	0	3,360	0	3,360
Kansas	416	21.6	3,710	111	5.60	53.3	0	31.4	0	2,260	0	6,610	0	6,610
Kentucky	525	27.5	29.3	—	—	317	0	—	—	3,260	0	4,160	0	4,160
Louisiana	753	41.2	1,020	7.34	243	2,680	0	—	—	5,610	0	10,400	0	10,400
Maine	102	35.7	5.84	—	—	247	0	—	—	113	295	504	295	799
Maryland	824	77.1	42.4	10.4	19.6	65.8	227	8.31	.02	379	6,260	1,430	6,490	7,910
Massachusetts	739	42.2	126	—	—	36.8	0	—	—	108	3,610	1,050	3,610	4,660
Michigan	1,140	239	201	11.3	—	698	0	—	—	7,710	0	10,000	0	10,000
Minnesota	500	80.8	227	52.8	—	154	0	588	0	2,270	0	3,870	0	3,870
Mississippi	359	69.3	1,410	—	371	242	0	—	—	362	148	2,810	148	2,960
Missouri	872	53.6	1,430	72.4	83.3	62.7	0	16.9	0	5,640	0	8,230	0	8,230
Montana	149	18.6	7,950	—	—	61.3	0	—	—	110	0	8,290	0	8,290
Nebraska	330	48.4	8,790	93.4	—	38.1	0	128	4.55	2,820	0	12,200	4.55	12,300
Nevada	629	22.4	2,110	—	—	10.3	0	—	—	36.7	0	2,810	0	2,810
New Hampshire	97.1	41.0	4.75	—	16.3	44.9	0	6.8	0	236	761	447	761	1,210
New Jersey	1,050	79.7	140	1.68	6.46	132	0	110	0	650	3,390	2,170	3,390	5,560
New Mexico	296	31.4	2,860	—	—	10.5	0	—	—	56.4	0	3,260	0	3,260
New York	2,570	142	35.5	—	—	297	0	—	—	4,040	5,010	7,080	5,010	12,100
North Carolina	945	189	287	121	7.88	293	0	36.4	0	7,850	1,620	9,730	1,620	11,400
North Dakota	63.6	11.9	145	—	—	17.6	0	—	—	902	0	1,140	0	1,140
Ohio	1,470	134	31.7	25.3	1.36	807	0	88.5	0	8,590	0	11,100	0	11,100
Oklahoma	675	25.5	718	151	16.4	25.9	0	2.48	256	146	0	1,760	256	2,020
Oregon	566	76.2	6,080	—	—	195	0	—	—	15.3	0	6,930	0	6,930
Pennsylvania	1,460	132	13.9	—	—	1,190	0	182	0	6,980	0	9,950	0	9,950
Rhode Island	119	8.99	3.45	—	—	4.28	0	—	—	2.40	290	138	290	429
South Carolina	566	63.5	267	—	—	565	0	—	—	5,710	0	7,170	0	7,170
South Dakota	93.3	9.53	373	42.0	—	5.12	0	—	—	5.24	0	528	0	528
Tennessee	890	32.6	22.4	—	—	842	0	—	—	9,040	0	10,800	0	10,800
Texas	4,230	131	8,630	308	—	1,450	907	220	504	9,820	3,440	24,800	4,850	29,600
Utah	638	16.1	3,860	—	116	42.7	5.08	26.3	198	62.2	0	4,760	203	4,970
Vermont	60.1	21.0	3.78	—	—	6.91	0	—	—	355	0	447	0	447
Virginia	720	133	26.4	—	—	470	53.3	—	—	3,850	3,580	5,200	3,640	8,830
Washington	1,020	125	3,040	—	—	577	39.9	—	—	519	0	5,270	39.9	5,310
West Virginia	190	40.4	.04	—	—	968	0	—	—	3,950	0	5,150	0	5,150
Wisconsin	623	96.3	196	66.3	70.2	447	0	—	—	6,090	0	7,590	0	7,590
Wyoming	107	6.57	4,500	—	—	5.78	0	79.5	222	243	0	4,940	222	5,170
Puerto Rico	513	.88	94.5	—	—	11.2	0	—	—	0	2,190	620	2,190	2,810
U.S. Virgin Islands	6.09	1.69	.50	—	—	3.34	0	—	—	0	136	11.6	136	148
Total	**43,300**	**3,590**	**137,000**	**1,760**	**3,700**	**18,500**	**1,280**	**2,010**	**1,490**	**136,000**	**59,500**	**345,000**	**62,300**	**408,000**

SOURCE: Susan S. Hutson et al., "Table 2. Total Water Withdrawals by Water-Use Category, 2000," in *Estimated Use of Water in the United States in 2000*, U.S. Department of the Interior, U.S. Geological Survey, 2004, http://pubs.usgs.gov/circ/2004//circ1268/pdf/circular1268.pdf (accessed January 2, 2007) and revision data February 7, 2005, http://pubs.usgs.gov/circ/2004//circ1268/control/revisions.html (accessed January 2, 2007)

watering vegetation in parks, supplying public buildings, firefighting, and street washing. In 2000 water utility companies supplied 43.3 Bgal/d. (See Table 2.2.) The rest of the water shown in Table 2.2 under the various other categories was self-supplied. That is, the water was withdrawn from groundwater or surface water sources by the users, not by water utility companies.

According to Hutson et al., public suppliers serviced about 242 million people in 2000 (about 85% of the total U.S. population of 285 million at that time). (See Table 2.3.)

TABLE 2.3

Public supply water use, 2000

[Figures may not sum to totals because of independent rounding]

State	Population (in thousands) Total	Served by public supply Population	Population (in percent)	Withdrawals (in million gallons per day) By source Ground-water	Surface water	Total	Withdrawals (in thousand acre-feet per year) By source Ground-water	Surface water	Total
Alabama	4,450	3,580	80	281	553	834	315	620	935
Alaska	627	421	67	29.3	50.7	80.0	32.9	56.9	89.7
Arizona	5,130	4,870	95	469	613	1,080	526	688	1,210
Arkansas	2,670	2,320	87	132	289	421	148	324	472
California	33,900	30,100	89	2,800	3,320	6,120	3,140	3,730	6,860
Colorado	4,300	3,750	87	53.7	846	899	60.2	948	1,010
Connecticut	3,410	2,660	78	66.0	358	424	74.0	402	476
Delaware	784	617	79	45.0	49.8	94.9	50.5	55.9	106
District of Columbia	572	572	100	0	0	0	0	0	0
Florida	16,000	14,000	88	2,200	237	2,440	2,470	266	2,730
Georgia	8,190	6,730	82	278	968	1,250	311	1,090	1,400
Hawaii	1,210	1,140	94	243	7.60	250	272	8.52	281
Idaho	1,290	928	72	219	25.3	244	245	28.3	274
Illinois	12,400	10,900	88	353	1,410	1,760	396	1,580	1,970
Indiana	6,080	4,480	74	345	326	670	386	365	751
Iowa	2,930	2,410	83	303	79.8	383	340	89.5	429
Kansas	2,690	2,500	93	172	244	416	193	273	466
Kentucky	4,040	3,490	86	71.0	455	525	79.5	510	589
Louisiana	4,470	3,950	88	349	404	753	392	453	844
Maine	1,270	726	57	29.6	72.5	102	33.2	81.3	115
Maryland	5,300	4,360	82	84.6	740	824	94.8	829	924
Massachusetts	6,350	5,880	93	197	542	739	220	608	828
Michigan	9,940	7,170	72	247	896	1,140	277	1,000	1,280
Minnesota	4,920	3,770	77	329	171	500	369	192	561
Mississippi	2,840	2,190	77	319	40.4	359	357	45.3	402
Missouri	5,600	4,770	85	278	594	872	311	666	978
Montana	902	664	74	56.1	92.4	149	62.9	104	167
Nebraska	1,710	1,390	81	266	63.8	330	299	71.6	370
Nevada	2,000	1,870	94	151	478	629	169	536	705
New Hampshire	1,240	756	61	33.0	64.1	97.1	37.0	71.9	109
New Jersey	8,410	7,460	89	400	650	1,050	449	729	1,180
New Mexico	1,820	1,460	80	262	33.8	296	294	37.9	332
New York	19,000	17,100	90	583	1,980	2,570	653	2,220	2,880
North Carolina	8,050	5,350	66	166	779	945	186	873	1,060
North Dakota	642	493	77	32.4	31.2	63.6	36.3	35.0	71.3
Ohio	11,400	9,570	84	500	966	1,470	560	1,080	1,640
Oklahoma	3,450	3,150	91	113	562	675	127	631	757
Oregon	3,420	2,730	80	118	447	566	133	501	634
Pennsylvania	12,300	10,100	82	212	1,250	1,460	237	1,400	1,640
Rhode Island	1,050	922	88	16.9	102	119	19.0	115	134
South Carolina	4,010	3,160	79	105	462	566	117	517	635
South Dakota	755	625	83	54.2	39.1	93.3	60.7	43.9	105
Tennessee	5,690	5,240	92	321	569	890	360	638	997
Texas	20,900	19,700	94	1,260	2,970	4,230	1,420	3,330	4,740
Utah	2,230	2,180	97	364	274	638	408	307	715
Vermont	609	362	59	19.5	40.6	60.1	21.8	45.6	67.4
Virginia	7,080	5,310	75	70.7	650	720	79.3	728	808
Washington	5,890	4,900	83	464	552	1,020	520	619	1,140
West Virginia	1,810	1,300	72	41.6	149	190	46.6	167	213
Wisconsin	5,360	3,620	67	330	293	623	370	329	699
Wyoming	494	406	82	57.2	49.4	107	64.1	55.3	119
Puerto Rico	3,810	3,800	100	88.5	425	513	99.2	476	576
U.S. Virgin Islands	109	53.4	49	.52	5.57	6.09	.58	6.24	6.83
Total	**285,000**	**242,000**	**85**	**16,000**	**27,300**	**43,300**	**17,900**	**30,600**	**48,500**

SOURCE: Susan S. Hutson et al., "Table 5. Public-Supply Water Withdrawals, 2000," in *Estimated Use of Water in the United States in 2000*, U.S. Department of the Interior, U.S. Geological Survey, 2004, http://pubs.usgs.gov/circ/2004//circ1268/pdf/circular1268.pdf (accessed January 2, 2007) and revision data February 7, 2005, http://pubs.usgs.gov/circ/2004//circ1268/control/revisions.html (accessed January 2, 2007)

This figure represents an 8% increase over the number of people supplied with water by public suppliers in 1995. Of the 43.3 Bgal/d of water that public suppliers withdrew, 27.3 Bgal/d (63%) came from surface sources and 16 Bgal/d (37%) from groundwater sources. Figure 2.3 shows that California, Texas, Illinois, New York, and Florida accounted for a majority of U.S. public-supply withdrawals in 2000.

DOMESTIC USE. Domestic water use includes water for typical household purposes, such as drinking; food preparation; bathing; washing clothes, dishes, and cars;

FIGURE 2.3

Public supply water use by source and state, 2000

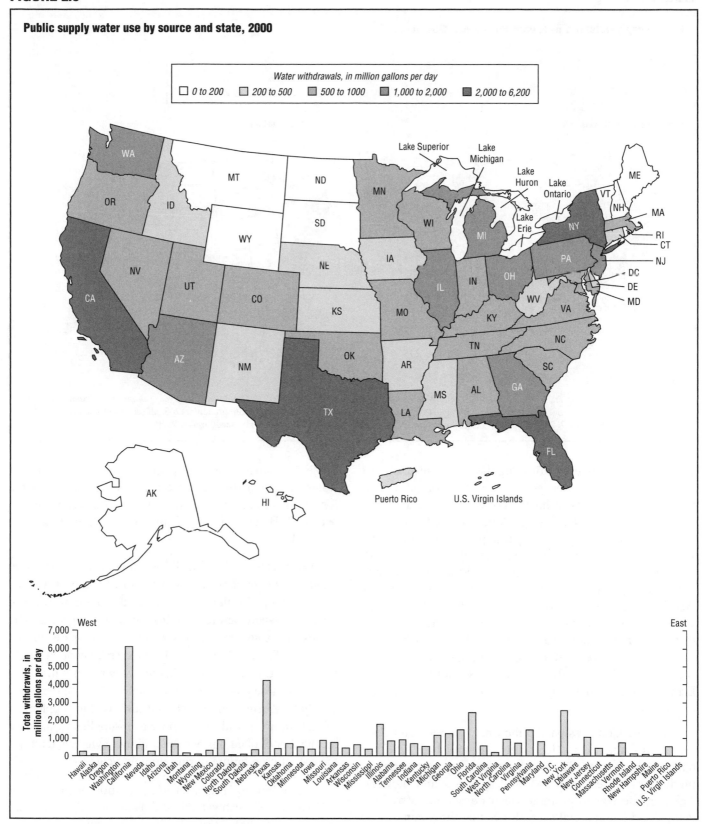

flushing toilets; and watering lawns and gardens. Although people need to take in about two quarts of water a day from what they eat or drink to replace water loss, water needs for household use (indoor and outdoor) add to the amount of water people require each day.

In the report *Residential End Uses of Water Study* (1999, http://www.awwarf.org/research/topicsandprojects/execSum/241.aspx), Peter W. Mayer et al. studied residential end uses of water in twelve hundred single-family homes in twelve North American locations from 1996 to

FIGURE 2.3

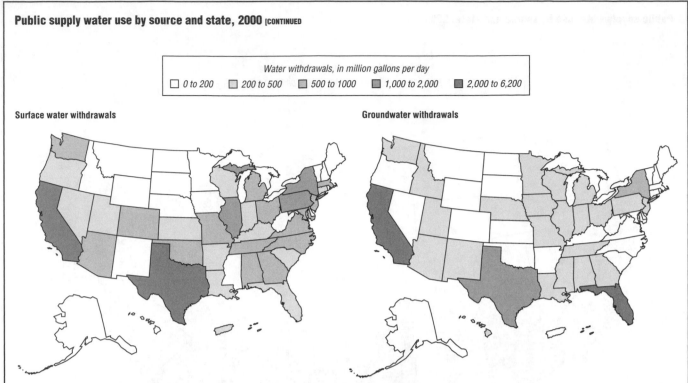

Public supply water use by source and state, 2000 [CONTINUED]

Water withdrawals, in million gallons per day

☐ 0 to 200   ☐ 200 to 500   ☐ 500 to 1000   ☐ 1,000 to 2,000   ■ 2,000 to 6,200

Surface water withdrawals

Groundwater withdrawals

SOURCE: Susan S. Hutson et al., "Figure 5. Public-Supply Withdrawals by Source and State, 2000," in *Estimated Use of Water in the United States in 2000*, U.S. Department of the Interior, U.S. Geological Survey, 2004, http://pubs.usgs.gov/circ/2004//circ1268/pdf/circular1268.pdf (accessed January 2, 2007) and revision data February 7, 2005, http://pubs.usgs.gov/circ/2004//circ1268/control/revisions.html (accessed January 2, 2007)

1998. They conclude that flushing the toilet uses the most water (20.1 gallons per person per day) of all indoor household water uses in homes not equipped with water-efficient fixtures. Laundering clothes ranked second in water use (15 gallons per person per day), and taking showers (13.3 gallons per person per day) ranked third. In addition, the U.S. Environmental Protection Agency (EPA) reports in the fact sheet "Safe Drinking Water Act 30th Anniversary: Water Facts" (June 2004, http://www.epa.gov/ogwdw/sdwa/30th/factsheets/waterfacts.html) that outdoor household water use for activities such as filling swimming pools, watering lawns, and washing cars accounts for 50% to 70% of total household water usage. Each American, on average, uses over one hundred gallons of water per day.

In 1992 Congress passed the Energy Policy and Conservation Act. This legislation established uniform national standards for manufacture of water-efficient plumbing fixtures, such as low-flow toilets and showers. The purpose was to promote water conservation by residential and commercial users. Since that time many water suppliers have sponsored programs offering rebates on water bills and other incentives to encourage the use of these devices to reduce water use.

AQUACULTURE. Aquaculture is the practice of raising animals that live in water—such as finfish and shellfish—for food, restoration, conservation, or sport. According to

Table 2.4, aquaculture use accounted for 3.7 Bgal/d of water use nationally in 2000. Surface water was the source for 2.6 Bgal/d (70%) of this total. Idaho alone accounted for nearly 2 Bgal/d (53%) of the aquaculture water use reported.

IRRIGATION. The word *irrigation* usually brings to mind arid or semiarid deserts transformed into lush green fields of crops by the turn of a handle, bringing life and prosperity where before there had been only sagebrush and cactus. To some extent this is true. Many parts of the American West and Midwest do not average enough yearly rainfall to sustain the crops that are grown there; the cultivation of those crops is made possible only with the water supplied by irrigation. Irrigation is also used to supplement rainfall in areas with adequate water supplies to increase the number of plantings per year, improve yield, and reduce the risk of crop failure during drought years.

According to Hutson et al., irrigation accounted for 137 Bgal/d of freshwater withdrawals for all off-stream categories in 2000. (See Table 2.2.) Approximately 80 Bgal/d of withdrawals (58%) were from surface water sources, and the remaining 56.9 Bgal/d (42%) were from groundwater sources. (See Table 2.5.) The quantity of freshwater used for irrigation varies greatly from region to region. Irrigation is by far the largest water use category in the West. (See Figure 2.4.) California alone used

TABLE 2.4

Water use for aquaculture, 2000

[Figures may not sum to totals because of independent rounding]

State	Withdrawals (in million gallons per day)			Withdrawals (in thousand acre-feet per year)		
	By source			By source		
	Groundwater	Surface water	Total	Groundwater	Surface water	Total
Alabama	8.93	1.44	10.4	10.0	1.61	11.6
Alaska	—	—	—	—	—	—
Arizona	—	—	—	—	—	—
Arkansas	187	10.4	198	210	11.6	222
California	158	380	537	177	426	603
Colorado	—	—	—	—	—	—
Connecticut	—	—	—	—	—	—
Delaware	.07	0	.07	.08	0	.08
District of Columbia	—	—	—	—	—	—
Florida	7.81	.21	8.02	8.76	.24	8.99
Georgia	7.70	7.72	15.4	8.63	8.65	17.3
Hawaii	—	—	—	—	—	—
Idaho	51.5	1,920	1,970	57.7	2,150	2,210
Illinois	—	—	—	—	—	—
Indiana	—	—	—	—	—	—
Iowa	—	—	—	—	—	—
Kansas	3.33	2.27	5.60	3.73	2.54	6.28
Kentucky	—	—	—	—	—	—
Louisiana	128	115	243	144	129	273
Maine	—	—	—	—	—	—
Maryland	4.81	14.8	19.6	5.39	16.6	22.0
Massachusetts	—	—	—	—	—	—
Michigan	—	—	—	—	—	—
Minnesota	—	—	—	—	—	—
Mississippi	321	49.8	371	360	55.9	416
Missouri	2.01	81.3	83.3	2.25	91.2	93.4
Montana	—	—	—	—	—	—
Nebraska	—	—	—	—	—	—
Nevada	—	—	—	—	—	—
New Hampshire	3.12	13.1	16.3	3.50	14.7	18.2
New Jersey	6.46	0	6.46	7.24	0	7.24
New Mexico	—	—	—	—	—	—
New York	—	—	—	—	—	—
North Carolina	7.88	0	7.88	8.83	0	8.83
North Dakota	—	—	—	—	—	—
Ohio	1.36	0	1.36	1.52	0	1.52
Oklahoma	.29	16.1	16.4	.33	18.1	18.4
Oregon	—	—	—	—	—	—
Pennsylvania	—	—	—	—	—	—
Rhode Island	—	—	—	—	—	—
South Carolina	—	—	—	—	—	—
South Dakota	—	—	—	—	—	—
Tennessee	—	—	—	—	—	—
Texas	—	—	—	—	—	—
Utah	116	0	116	130	0	130
Vermont	—	—	—	—	—	—
Virginia	—	—	—	—	—	—
Washington	—	—	—	—	—	—
West Virginia	—	—	—	—	—	—
Wisconsin	39.8	30.4	70.2	44.6	34.1	78.7
Wyoming	—	—	—	—	—	—
Puerto Rico	—	—	—	—	—	—
U.S. Virgin Islands	—	—	—	—	—	—
Total	**1,060**	**2,640**	**3,700**	**1,180**	**2,960**	**4,150**

SOURCE: Susan S. Hutson et al., "Table 9. Aquaculture Water Withdrawals, 2000," in *Estimated Use of Water in the United States in 2000*, U.S. Department of the Interior, U.S. Geological Survey, 2004, http://pubs.usgs.gov/circ/2004//circ1268/pdf/circular1268.pdf (accessed January 2, 2007) and revision data February 7, 2005, http://pubs.usgs.gov/circ/2004//circ1268/control/revisions.html (accessed January 2, 2007)

30.5 Bgal/d (22%) of all irrigation water in 2000. (See Table 2.5.)

Hutson et al. note that irrigation has the highest consumptive use of any of the eight categories of off-stream water use. In many irrigated areas about 75% to 85% of the irrigation water is lost to evaporation, transpiration, or retained in the crops. The remaining 15% to 25% either slowly makes its way through the soil to recharge (replenish) groundwater, a process called irrigation return flow, or is returned to nearby surface water through a drainage system. The average quantities of

TABLE 2.5

Acres of land being irrigated by state, 2000

[Figures may not sum to totals because of independent rounding]

State	Irrigated land (in thousand acres) By type of irrigation Sprinkler	Micro-irrigation	Surface	Total	Withdrawals (in million gallons per day) By source Groundwater	Surface water	Total
Alabama	68.7	1.30	0	70.0	14.5	28.7	43.1
Alaska	2.43	0	.07	2.50	.99	.02	1.01
Arizona	183	14.0	779	976	2,750	2,660	5,400
Arkansas	631	0	3,880	4,510	6,510	1,410	7,910
California	1,660	3,010	5,470	10,100	11,600	18,900	30,500
Colorado	1,190	1.16	2,220	3,400	2,160	9,260	11,400
Connecticut	20.6	.39	0	21.0	17.0	13.4	30.4
Delaware	81.1	.71	0	81.8	35.6	7.89	43.5
District of Columbia	.32	0	0	.32	0	.18	.18
Florida	515	704	839	2,060	2,180	2,110	4,290
Georgia	1,470	73.8	0	1,540	750	392	1,140
Hawaii	16.7	105	0	122	171	193	364
Idaho	2,440	4.70	1,300	3,750	3,720	13,300	17,100
Illinois	365	0	0	365	150	4.25	154
Indiana	250	0	0	250	55.5	45.4	101
Iowa	84.5	0	0	84.5	20.4	1.08	21.5
Kansas	2,660	2.14	647	3,310	3,430	288	3,710
Kentucky	66.6	0	0	66.6	1.14	28.2	29.3
Louisiana	110	0	830	940	791	232	1,020
Maine	35.0	.95	.03	36.0	.61	5.23	5.84
Maryland	57.3	3.32	0	60.6	29.8	12.6	42.4
Massachusetts	26.6	2.35	0	29.0	19.7	106	126
Michigan	401	8.67	4.87	415	128	73.2	201
Minnesota	546	0	26.9	573	190	36.6	227
Mississippi	455	0	966	1,420	1,310	99.1	1,410
Missouri	532	1.43	792	1,330	1,380	48.1	1,430
Montana	506	0	1,220	1,720	83.0	7,870	7,950
Nebraska	4,110	0	3,710	7,820	7,420	1,370	8,790
Nevada	192	0	456	647	567	1,540	2,110
New Hampshire	6.08	0		6.08	.50	4.25	4.75
New Jersey	109	15.7	3.70	128	22.8	117	140
New Mexico	461	7.17	530	998	1,230	1,630	2,860
New York	70.0	8.73	1.84	80.6	23.3	12.1	35.5
North Carolina	193	3.70	0	196	65.8	221	287
North Dakota	200	0	26.7	227	72.2	73.2	145
Ohio	61.0	0	0	61.0	13.9	17.8	31.7
Oklahoma	392	1.50	113	507	566	151	718
Oregon	1,160	4.02	1,000	2,170	792	5,290	6,080
Pennsylvania	28.9	7.17	0	36.0	1.38	12.5	13.9
Rhode Island	4.48	.29	.05	4.82	.46	2.99	3.45
South Carolina	166	3.66	17.5	187	106	162	267
South Dakota	276	0	78.3	354	137	236	373
Tennessee	51.2	5.35	3.96	60.5	7.33	15.1	22.4
Texas	4,010	89.4	2,390	6,490	6,500	2,130	8,630
Utah	526	1.68	880	1,410	469	3,390	3,860
Vermont	4.95	0	0	4.95	.33	3.45	3.78
Virginia	64.3	13.9	0	78.2	3.57	22.8	26.4
Washington	1,270	49.9	252	1,570	747	2,290	3,040
West Virginia	2.21	0	.98	3.19	.02	.02	.04
Wisconsin	355	0	0	355	195	1.57	196
Wyoming	190	4.73	964	1,160	413	4,090	4,500
Puerto Rico	15.5	33.0	5.35	53.8	36.9	57.5	94.5
U.S. Virgin Islands	.20	0	0	.20	.29	.21	.50
Total	**28,300**	**4,180**	**29,400**	**61,900**	**56,900**	**80,000**	**137,000**

SOURCE: Adapted from Susan S. Hutson et al., "Table 7. Irrigation Water Withdrawals, 2000," in *Estimated Use of Water in the United States in 2000*, U.S. Department of the Interior, U.S. Geological Survey, 2004, http://pubs.usgs.gov/circ/2004//circ1268/pdf/circular1268.pdf (accessed January 2, 2007) and revision data February 7, 2005, http://pubs.usgs.gov/circ/2004//circ1268/control/revisions.html (accessed January 2, 2007)

water applied range from several inches to more than twenty inches per year, depending on local conditions.

Significant changes in water quality can be caused by irrigation. The water lost in evapotranspiration is relatively pure because nonwater chemicals are left behind, precipitating as salts and accumulating in the soil. The salts continue to accumulate as irrigation continues. Accumulation of salts in the soil can cause the concentration of salts in the irrigation return flows to be higher than in the original irrigation water. Excessive salts in the

FIGURE 2.4

Water use for irrigation by source and state, 2000

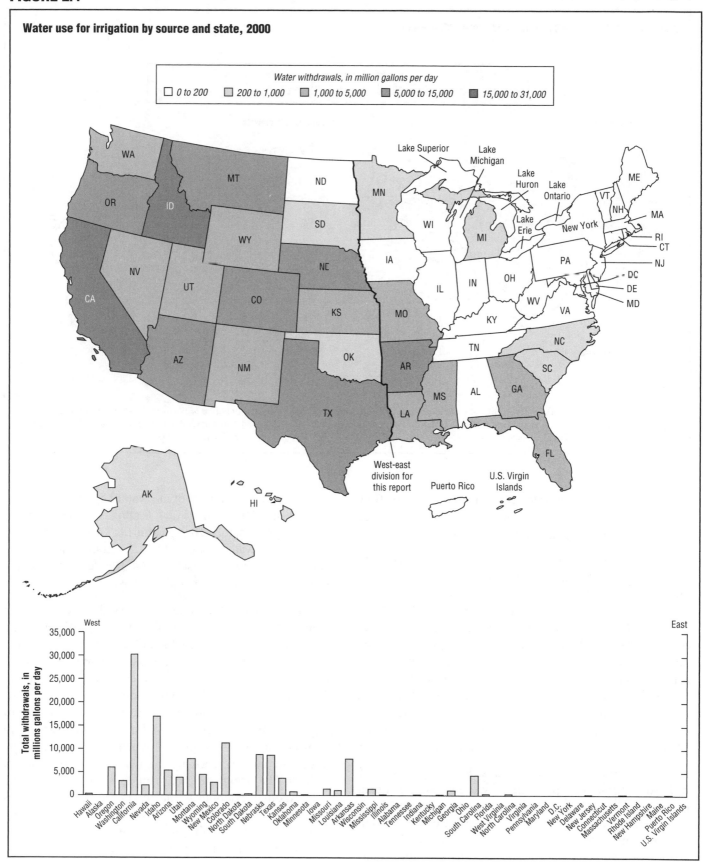

soil can also interfere with crop growth, sometimes resulting in soil unsuitable for crop growth. To stop excessive buildup of salts in the soil, extra irrigation water is often used to flush the salts from the soil and transport them into the groundwater. In locations where these dissolved salts reach high concentrations, the

FIGURE 2.4

Water use for irrigation by source and state, 2000 [CONTINUED]

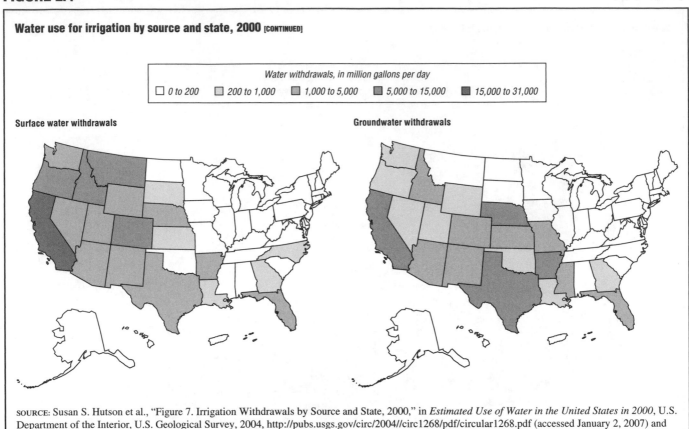

SOURCE: Susan S. Hutson et al., "Figure 7. Irrigation Withdrawals by Source and State, 2000," in *Estimated Use of Water in the United States in 2000*, U.S. Department of the Interior, U.S. Geological Survey, 2004, http://pubs.usgs.gov/circ/2004//circ1268/pdf/circular1268.pdf (accessed January 2, 2007) and revision data February 7, 2005, http://pubs.usgs.gov/circ/2004//circ1268/control/revisions.html (accessed January 2, 2007)

recharge of the groundwater from irrigation return flow can reduce the quality of groundwater and the surface water to which the groundwater discharges.

LIVESTOCK. Livestock water use includes drinking water for livestock, dairy and feedlot operations, and other on-farm needs. Hutson et al. indicate that 1.8 Bgal/d of water was used for these purposes in 2000. (See Table 2.2.) Total withdrawals for livestock increased slightly between 1995 and 2000 for the twenty-two states that reported data for both years. However, withdrawals actually increased only in eight of the twenty-two reporting states. Combined, California, Texas, and Oklahoma accounted for 868 million gallons per day (Mgal/d; 49%) of the U.S. total livestock water use in 2000, and 556 Mgal/d (66%) of the surface water used. (See Table 2.6.)

INDUSTRIAL. Even those industries that do not use water directly in their products may use substantial quantities of water during operations. Water for industrial use is commonly divided into four categories: cooling water, process water, boiler feed water, and sanitary and service water (for personal use by employees, for cleaning plants and equipment, and for the operation of valves and other equipment). Industries that use the most water include steel, chemical and allied products, paper and related products, and petroleum refining.

According to Hutson et al., water supplied for industrial use in 2000 totaled 19.7 Bgal/d, 11% less than in 1995. Approximately 16.2 Bgal/d (82%) was withdrawn from surface water. (See Table 2.7.) Louisiana, Indiana, and Texas together consumed 7.4 Bgal/d (37%) of the nation's industrial water withdrawals.

Most manufacturers use processed water at some point in the course of making a product. Water is the solvent (the substance in which other substances are dissolved) in many chemical processes. In some plants the item being manufactured is in contact with water at almost every step in its conversion from raw materials to finished product. For example, in the production of pulp and paper, water is used for removing bark from pulpwood, moving the ground wood and pulp from one process to another, cooking the wood chips for removal of lignin (the woody pulp of plant cells), and washing the pulp. Another example is the food industry, which uses huge quantities of water for cleaning and cooking vegetables and meat, canning and cooling canned products, and cleaning equipment and facilities.

The need for large quantities of easily accessible water has led to industrial development around or near coastlines, rivers, and lakes. The Great Lakes region and the Ohio River Valley are examples. This development

TABLE 2.6

Water use for livestock by state, 2000

[Figures may not sum to totals because of independent rounding]

State	Withdrawals (in million gallons per day) By source			Withdrawals (in thousand acre-feet per year) By source		
	Groundwater	Surface water	Total	Groundwater	Surface water	Total
Alabama	—	—	—	—	—	—
Alaska	—	—	—	—	—	—
Arizona	—	—	—	—	—	—
Arkansas	—	—	—	—	—	—
California	182	227	409	204	255	458
Colorado	—	—	—	—	—	—
Connecticut	—	—	—	—	—	—
Delaware	3.70	.22	3.92	4.15	.25	4.39
District of Columbia	—	—	—	—	—	—
Florida	31.0	1.51	32.5	34.7	1.69	36.4
Georgia	1.66	17.7	19.4	1.86	19.9	21.7
Hawaii	—	—	—	—	—	—
Idaho	27.7	7.20	34.9	31.0	8.07	39.1
Illinois	37.6	0	37.6	42.1	0	42.1
Indiana	27.3	14.6	41.9	30.6	16.4	47.0
Iowa	81.8	27.1	109	91.8	30.4	122
Kansas	87.2	23.5	111	97.7	26.3	124
Kentucky	—	—	—	—	—	—
Louisiana	4.03	3.31	7.34	4.52	3.71	8.23
Maine	—	—	—	—	—	—
Maryland	7.18	3.18	10.4	8.05	3.56	11.6
Massachusetts	—	—	—	—	—	—
Michigan	10.2	1.15	11.3	11.4	1.29	12.7
Minnesota	52.8	0	52.8	59.2	0	59.2
Mississippi	—	—	—	—	—	—
Missouri	18.3	54.1	72.4	20.5	60.6	81.1
Montana	—	—	—	—	—	—
Nebraska	76.0	17.4	93.4	85.2	19.5	105
Nevada	—	—	—	—	—	—
New Hampshire	—	—	—	—	—	—
New Jersey	1.68	0	1.68	1.88	0	1.88
New Mexico	—	—	—	—	—	—
New York	—	—	—	—	—	—
North Carolina	89.1	32.3	121	99.9	36.2	136
North Dakota	—	—	—	—	—	—
Ohio	8.20	17.1	25.3	9.19	19.2	28.4
Oklahoma	53.6	97.2	151	60.0	109	169
Oregon	—	—	—	—	—	—
Pennsylvania	—	—	—	—	—	—
Rhode Island	—	—	—	—	—	—
South Carolina	—	—	—	—	—	—
South Dakota	16.9	25.2	42.0	18.9	28.2	47.1
Tennessee	—	—	—	—	—	—
Texas	137	172	308	153	192	346
Utah	—	—	—	—	—	—
Vermont	—	—	—	—	—	—
Virginia	—	—	—	—	—	—
Washington	—	—	—	—	—	—
West Virginia	—	—	—	—	—	—
Wisconsin	60.3	6.02	66.3	67.6	6.75	74.4
Wyoming	—	—	—	—	—	—
Puerto Rico	—	—	—	—	—	—
U.S. Virgin Islands	—	—	—	—	—	—
Total	1,010	747	1,760	1,140	838	1,980

SOURCE: Susan S. Hutson et al., "Table 8. Livestock Water Withdrawals, 2000," in *Estimated Use of Water in the United States in 2000*, U.S. Department of the Interior, U.S. Geological Survey, 2004, http://pubs.usgs.gov/circ/2004//circ1268/pdf/circular1268.pdf (accessed January 2, 2007) and revision data February 7, 2005, http://pubs.usgs.gov/circ/2004//circ1268/control/revisions.html (accessed January 2, 2007)

has often caused serious deterioration of water quality in the area because, after it is used, water may be returned to its source carrying pollutants.

MINING. Mining is the extraction of naturally occurring materials, including petroleum, from the earth's crust. Water is used for washing and milling (processing). All water for mining operations is self-supplied and may come from a freshwater or saline source. Hutson et al. classify water as saline if it contains more than one thousand milligrams per liter or more of dissolved solids (salts).

TABLE 2.7

Industrial water use by state, 2000

[Figures may not sum to totals because of independent rounding]

State	Withdrawals (in million gallons per day) By source and type Groundwater			Surface water			Total			Withdrawals (in thousand acre-feet per year) By type		
	Fresh	Saline	Total	Fresh	Saline	Total	Fresh	Saline	Total	Fresh	Saline	Total
Alabama	56.0	0	56.0	777	0	777	833	0	833	934	0	934
Alaska	4.32	0	4.32	3.80	3.86	7.66	8.12	3.86	12.0	9.10	4.33	13.4
Arizona	19.8	0	19.8	0	0	0	19.8	0	19.8	22.2	0	22.2
Arkansas	67.0	.08	67.1	66.8	0	66.8	134	.08	134	150	.09	150
California	183	0	183	5.65	13.6	19.3	188	13.6	202	211	15.3	226
Colorado	23.6	0	23.6	96.4	0	96.4	120	0	120	135	0	135
Connecticut	4.13	0	4.13	6.61	0	6.61	10.7	0	10.7	12.0	0	12.0
Delaware	17.0	0	17.0	42.5	3.25	45.7	59.4	3.25	62.7	66.6	3.64	70.3
District of Columbia	0	0	0	0	0	0	0	0	0	0	0	0
Florida	216	0	216	74.7	1.18	75.9	291	1.18	292	326	1.32	328
Georgia	290	0	290	333	30	363	622	30.0	652	698	33.6	731
Hawaii	14.5	.85	15.4	0	0	0	14.5	.85	15.4	16.2	0.95	17.2
Idaho	35.8	0	35.8	19.7	0	19.7	55.5	0	55.5	62.2	0	62.2
Illinois	132	0	132	259	0	259	391	0	391	438	0	438
Indiana	99.7	0	99.7	2,300	0	2,300	2,400	0	2,400	2,690	0	2,690
Iowa	226	0	226	11.7	0	11.7	237	0	237	266	0	266
Kansas	46.6	0	46.6	6.74	0	6.74	53.3	0	53.3	59.8	0	59.8
Kentucky	95.2	0	95.2	222	0	222	317	0	317	356	0	356
Louisiana	285	0	285	2,400	0	2,400	2,680	0	2,680	3,010	0	3,010
Maine	9.90	0	9.90	237	0	237	247	0	247	277	0	277
Maryland	15.9	0	15.9	49.9	227	277	65.8	227	292	73.8	254	328
Massachusetts	10.7	0	10.7	26.2	0	26.2	36.8	0	36.8	41.3	0	41.3
Michigan	110	0	110	589	0	589	698	0	698	782	0	782
Minnesota	56.3	0	56.3	97.8	0	97.8	154	0	154	173	0	173
Mississippi	118	0	118	124	0	124	242	0	242	271	0	271
Missouri	29.2	0	29.2	33.5	0	33.5	62.7	0	62.7	70.3	0	70.3
Montana	31.9	0	31.9	29.3	0	29.3	61.3	0	61.3	68.7	0	68.7
Nebraska	35.5	0	35.5	2.60	0	2.60	38.1	0	38.1	42.7	0	42.7
Nevada	5.29	0	5.29	5.00	0	5.00	10.3	0	10.3	11.5	0	11.5
New Hampshire	6.95	0	6.95	37.9	0	37.9	44.9	0	44.9	50.3	0	50.3
New Jersey	65.3	0	65.3	66.2	0	66.2	132	0	132	147	0	147
New Mexico	8.80	0	8.80	1.67	0	1.67	10.5	0	10.5	11.7	0	11.7
New York	145	0	145	152	0	152	297	0	297	333	0	333
North Carolina	25.6	0	25.6	267	0	267	293	0	293	329	0	329
North Dakota	6.88	0	6.88	10.7	0	10.7	17.6	0	17.6	19.7	0	19.7
Ohio	162	0	162	645	0	645	807	0	807	905	0	905
Oklahoma	6.83	0	6.83	19.1	0	19.1	25.9	0	25.9	29.1	0	29.1
Oregon	12.1	0	12.1	183	0	183	195	0	195	218	0	218
Pennsylvania	155	0	155	1,030	0	1,030	1,190	0	1,190	1,330	0	1,330
Rhode Island	2.19	0	2.19	2.09	0	2.09	4.28	0	4.28	4.80	0	4.80
South Carolina	50.9	0	50.9	514	0	514	565	0	565	633	0	633
South Dakota	3.16	0	3.16	1.96	0	1.96	5.12	0	5.12	5.74	0	5.74
Tennessee	56.3	0	56.3	785	0	785	842	0	842	944	0	944
Texas	244	.50	244	1,200	906	2,110	1,450	907	2,350	1,620	1,020	2,640
Utah	34.3	5.08	39.4	8.38	0	8.38	42.7	5.08	47.8	47.8	5.69	53.5
Vermont	2.05	0	2.05	4.86	0	4.86	6.91	0	6.91	7.75	0	7.75
Virginia	104	0	104	365	53.3	419	470	53.3	523	526	59.7	586
Washington	138	0	138	439	39.9	479	577	39.9	617	647	44.7	692
West Virginia	9.70	0	9.70	958	0	958	968	0	968	1,090	0	1,090
Wisconsin	83.0	0	83.0	364	0	364	447	0	447	501	0	501
Wyoming	4.31	0	4.31	1.47	0	1.47	5.78	0	5.78	6.48	0	6.48
Puerto Rico	11.2	0	11.2	0	0	0	11.2	0	11.2	12.5	0	12.5
U.S. Virgin Islands	.22	0	.22	3.12	0	3.12	3.34	0	3.34	3.74	0	3.74
Total	3,570	6.51	3,580	14,900	1,280	16,200	18,500	1,280	19,700	20,700	1,440	22,100

SOURCE: Susan S. Hutson et al., "Table 10. Industrial Self-Supplied Water Withdrawals, 2000," in *Estimated Use of Water in the United States in 2000*, U.S. Department of the Interior, U.S. Geological Survey, 2004, http://pubs.usgs.gov/circ/2004//circ1268/pdf/circular1268.pdf (accessed January 2, 2007) and revision data February 7, 2005, http://pubs.usgs.gov/circ/2004//circ1268/control/revisions.html (accessed January 2, 2007)

Hutson et al. estimate that 3.5 Bgal/d of water was withdrawn for mining in 2000, down from 3.8 Bgal/d in 1995. Most water used for mining purposes was in the Texas Gulf area, followed by the Great Lakes region. Texas, Minnesota, and Wyoming together accounted for 1.6 Bgal/d (46%) of the mining withdrawals reported.

Acid mine drainage is a byproduct of mining activity. It is the drainage that results from the activity of removing

TABLE 2.8

Total water use by usage category, 2000

Category	Percent
Public supply	11
Irrigation	34
Aquaculture	<1
Mining	<1
Domestic	<1
Livestock	<1
Industrial	5
Thermoelectric power	48

SOURCE: Adapted from Susan S. Hutson et al., "Figure 1. Total Water Withdrawals by Category, 2000," in *Estimated Use of Water in the United States in 2000*, U.S. Department of the Interior, U.S. Geological Survey, 2004, http://pubs.usgs.gov/circ/2004//circ1268/pdf/circular1268.pdf (accessed January 2, 2007) and revision data February 7, 2005, http://pubs.usgs.gov/circ/2004//circ1268/control/revisions.html (accessed January 2, 2007)

and processing large amounts of rock to recover desired ores of heavy metals, minerals, and coal. Thousands of miles of streams are severely affected by drainage and runoff from abandoned coal mines, which are the single largest source of adverse water-quality impacts to both surface and groundwater in the United States.

THERMOELECTRIC POWER. Thermoelectric power plants are those that use turbines or similar devices to convert pressurized steam into electricity. Hydroelectric power plants use moving water to produce electricity, and nuclear power plants use water to cool nuclear reactors. Only the thermoelectric plants remove water for off-stream use. The water used in hydroelectric plants and nuclear power plants is in-stream use, so it is not included here.

Water used for thermoelectric power generation accounted for almost half (48%) of all withdrawals for off-stream use in 2000. (See Table 2.8.) Hutson et al. note that the largest total withdrawals were in Texas. States in the eastern portion of the country accounted for about 83% of the total thermoelectric water use. Figure 2.5 shows the geographic distribution of total, total freshwater, and total saline water withdrawals for thermoelectric power. These maps also visually show that California, Texas, and states in the eastern United States use the most off-stream water to produce electricity. By contrast, the Pacific Northwest uses hydroelectric power generation (an in-stream use) to supply a substantial part of the regional demand for electricity.

WASTE DISPOSAL. Water has been used to dilute and disperse waste since the earliest human settlements. If the wastewater is properly treated, the water environment can dilute, disperse, and assimilate waste products without harm to water quality or aquatic communities. The first step in the process is to identify the total maximum daily load of individual pollutants that particular water bodies can receive and not violate state water quality standards. The next step is to design, construct, and operate wastewater treatment facilities that provide the necessary level of treatment before discharging wastewater.

For the first (and last) time, the USGS reports in *Estimated Use of Water in the United States in 1995* (1998, http://water.usgs.gov/watuse/pdf1995/pdf/wastewater.pdf) the wastewater releases and return flow in 1995. This category includes facilities that collect, treat, and dispose of water through sewer systems and wastewater treatment plants, generally to surface waters. Over 16,400 publicly owned treatment facilities released 41 Bgal/d of treated wastewater nationwide in 1995. The annual average was one to two million gallons of treated water per facility per day. The largest wastewater return flows occurred in regions with large populations. Illinois (4.8 Bgal/d) and Ohio (4.7 Bgal/d) reported the largest releases of treated wastewater.

Not all treated wastewater is return flow. Because of the increasing demand for water and the cost of treating drinking water, more emphasis is being placed on water conservation and water reclamation (reuse). Reclaimed water is wastewater that has been treated for uses such as irrigation of golf courses or public parks instead of being discharged back to source waters. Florida (271 Mgal/d), California (216 Mgal/d), and Arizona (209 Mgal/d) reported large uses of reclaimed wastewater in 1995.

RIGHT TO WATER USE

The off-stream water-use categories described earlier are generally recognized as representing the most essential human uses of water. Sometimes there is not enough water available at a given location to meet all the demands for it. In these situations, who owns the water?

Water rights are held in trust by the states (held by the state for the benefit of the state's residents) and may be assigned to individuals and corporations according to statutes (laws) regulating water use. A state may also challenge water use to ensure public access to water that lies within or along its boundaries. State laws, regulations, and procedures establish how an individual, company, or other organization obtains and protects water rights. When water rights are disputed, particularly in the West, the question is often resolved through a judicial determination known as adjudication. According to the U.S. Fish and Wildlife Service (USFWS; April 29, 1993, http://www.fws.gov/policy/403fw2.html), adjudications may determine "all rights to use water in a particular stream system or watershed to establish the priority, point of diversion, place and nature of use, and the quantity of water used among the various claimants." When the water involved crosses state boundaries, states enter into agreements for water sharing. When agreement cannot be reached between states, the matter is usually settled in the federal courts, or in some cases by an act of Congress.

FIGURE 2.5

Water use (fresh and saline) for thermoelectric power by state, 2000

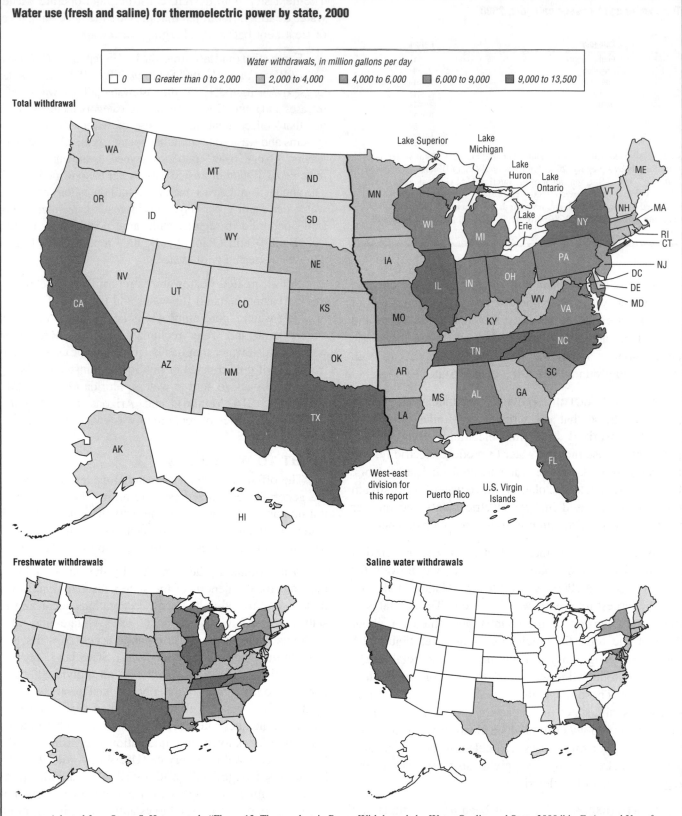

Water withdrawals, in million gallons per day

☐ 0   ☐ Greater than 0 to 2,000   ☐ 2,000 to 4,000   ☐ 4,000 to 6,000   ☐ 6,000 to 9,000   ■ 9,000 to 13,500

Total withdrawal

Freshwater withdrawals

Saline water withdrawals

SOURCE: Adapted from Susan S. Hutson et al., "Figure 12. Thermoelectric-Power Withdrawals by Water Quality and State, 2000," in *Estimated Use of Water in the United States in 2000*, U.S. Department of the Interior, U.S. Geological Survey, 2004, http://pubs.usgs.gov/circ/2004//circ1268/pdf/circular 1268.pdf (accessed January 2, 2007) and revision data February 7, 2005, http://pubs.usgs.gov/circ/2004//circ1268/control/revisions.html (accessed January 2, 2007)

Riparian Rights

The right of private landowners to use the water next to their property in streams, lakes, ponds, and other bodies of water is known as a riparian right, and this right underlies the laws regulating water in most states in the eastern part of the country. Even though local statutes are often written to pertain specifically to the bodies of water they regulate, the riparian system generally assigns each landowner an equal right to reasonable use of the water. Defining reasonable use can lead to disputes among neighboring landowners, but it typically allows common agricultural and private uses that do not involve holding water in storage.

Doctrine of Prior Appropriation

The relative scarcity of water in the West has led to a unique set of laws regulating water use, based on what is known as the doctrine of prior appropriation. Covering both surface and groundwater, the appropriation doctrine determines water rights by applying two standards: the timing of water claims and the nature of the water use. Rather than assigning rights based on landownership, the doctrine of appropriation considers both when and why water is used. The earliest water user is considered to hold a claim to the water, and the extent of those rights are judged by whether or not that use is considered beneficial. To determine beneficial use, two areas are considered: the purpose of the use and its efficiency (i.e., that the use is not wasteful). Individual states define the scope of beneficial uses within their boundaries.

A notable difference from the riparian (landownership) system is that under the doctrine of appropriation, water rights may be forfeited if the rights holder fails to use the water in a manner approved by the state or discontinues beneficial use for a designated length of time.

Another feature of the appropriation system is that rights are not shared equally among all users of a body of water. According to the USFWS, "Priority determines the order of rank of the rights to use water in a system. . . . [The] person first using water for a beneficial purpose has a right superior to those commencing their use later." Therefore, when water shortages occur, all rights holders are not affected equally as they are under the riparian system. Prior claims take precedence. However, because water shortages in the West can affect community water needs, priority may be awarded to some vital uses regardless of the date of first claim.

Conflicts over Federal Water Rights

Sometimes conflicts over water rights arise that involve the federal government, and since the 1990s the federal government has been involved in several long court battles to determine the precedence of water claims in the West. In *Tulare Lake Basin Water Storage District v. U.S. 49 Fed. Cl. 313* (2001), Judge John Wiese found that even though the federal government had a right to withhold water from farmers for irrigation to preserve salmon and smelt in California, by doing so the government had deprived farmers in the San Joaquin Valley of their rightful use of water. In December 2004 the Bush administration agreed to pay $16.7 million to compensate the farmers for their loss. The case was considered to have negative implications for environmental projects in the West, where the high costs associated with preservation efforts might make them untenable. Previously, the protection of endangered species was considered a higher priority than individual rights to water access.

In another case involving the federal government, *Trout Unlimited v. U.S. Department of Agriculture* (D.C. No. 96-WY-2686-WD), a conservation organization challenged the approval by the U.S. Forest Service of access to the Long Draw Reservoir in Colorado that did not establish bypass flow regulations for water projects. A bypass flow is the minimum amount of water needed to flow freely around a dam or diversion to sustain the area's aquatic life. On April 30, 2004, a federal judge determined that the Forest Service had not only the authority but also a responsibility to consider the protection of wildlife when issuing permits for water projects on federal lands.

PROTECTION OF AQUATIC LIFE

Since the enactment of the Clean Water Act in 1972, with its emphasis on maintaining the physical, chemical, and biological characteristics of the nation's waters, there has been an increasing awareness of the need to protect and maintain the insects, plants, and animals that make up the ecosystem of surface water bodies. Because life on Earth began in the ancient seas, aquatic life has been an integral part of overall water resources. This fact has frequently been ignored as human civilizations evolved, resulting in widespread change in and annihilation of aquatic systems.

In the United States, allocating water to maintain aquatic systems was rarely recognized as a legitimate use until the last two decades of the twentieth century. Before that time dam construction frequently disrupted whole ecological systems by reducing the water available to aquatic life in large stretches of rivers and streams below dams, interfering with the life cycles of migrating fish and other organisms and flooding habitats. In some river systems, such as the Colorado River, the entire flow was allocated and appropriated, resulting in drastic changes to the lush waterscape observed decades before at the delta of the Sea of Cortez, where the Colorado River deposited its rich silt. Rivers and streams have been lined with impermeable surfaces such as concrete or channelized to conserve water, control flooding, or provide passage for boats.

These practices are slowly changing. Permits issued for dam construction or reissued for dam operation are beginning to contain a provision for maintenance of minimum flow below the dam at a level sufficient to protect the natural system. In several cases this has required reduction in the water allocated to other users. Many states have programs to restore natural systems by removing abandoned or obsolete dams and other waterway obstructions and by constructing fish ladders to facilitate fish passage. Water allocation decisions in areas where water is a scarce resource are increasingly designating a portion for aquatic life protection. Proposals to divert or use water are more closely scrutinized to avoid adverse impacts to aquatic life. Recognizing aquatic life protection as a legitimate water use will have a profound effect on future water allocation decisions.

Except for a few rare instances, water is owned by the states, not the federal government. Therefore, the USFWS has adopted a policy of obtaining water rights. The objective is to obtain water supplies of adequate quantity and quality and the legal rights to use that water from the states, for development, use, and management of USFWS lands and facilities and for other congressionally authorized objectives, such as protection of endangered species and maintenance of in-stream flows.

The following are some examples of the evolving recognition of aquatic life protection as a legitimate water use:

- In May 2004 President George W. Bush issued an executive order that created a federal Great Lakes Interagency Task Force that would work to improve the deteriorating health of the Great Lakes. This task force and other regional groups convened in December 2004 and developed the *Great Lakes Regional Collaboration Strategy to Restore and Protect the Great Lakes* (December 12, 2005, http://www.glrc.us/documents/strategy/GLRC_Strategy.pdf). This strategy includes plans to stop the overflow of untreated sewage into the lakes, reduce agricultural runoff, protect wetlands, and control foreign species such as zebra mussels that are disrupting the aquatic food chain. As of March 2007, project activities were just beginning.

- A February 26, 2003, news report issued by the National Marine Fisheries Service stated that for the second consecutive year federal agencies had made substantial progress in implementing the National Marine Fisheries Service's 2000 Biological Opinion for the Federal Columbia River Power System. These efforts have resulted in the protection of hundreds of miles of habitat and in a record return of adult fish to the Columbia River in 2002.

- In "Region 9: Progress Report 2002" (February 26, 2007, http://www.epa.gov/region09/annualreport/02/

water.html), the EPA discusses the success of the salmon recovery project in Northern California's Butte Creek. The project, which was undertaken by the CALFED Bay-Delta Program, has resulted in an average spring salmon spawning of about six thousand fish—up from about one thousand fish per spring from the 1960s through the 1990s. The removal of four small dams that had blocked salmon passage was funded by the local Western Canal Water District and Southern California's Metropolitan Water District.

- According to the Connecticut River Coordinator's Office (March 26, 2007, http://www.fws.gov/r5crc/Habitat/fish_passage.htm) of the USFWS, removal of the New England Box Company dam on the Ashuelot River in Winchester, New Hampshire, in 2002 restored approximately fifteen miles of the river to free-flowing for the first time in one hundred years. The project was part of a river restoration plan intended to help bring back thousands of American shad, blueback herring, and Atlantic salmon to the river. As one of New Hampshire's major tributaries to the Connecticut River, the Ashuelot is historically important for migratory fish.

- In March 2000 Judge Richard Hicks of the Thurston County Superior Court ruled that the Washington Department of Ecology had to implement a 1993 statute requiring metering of water use throughout the state. The implementation had to include both surface and groundwater. The water metering statute was adopted as part of a larger salmon recovery package and was seen as an essential element in the wise management of the state's water resources for both people and salmon. Metering is viewed as an effective way to get the basic information about who is using the water and how much.

TRENDS IN WATER USE SINCE 1950

After continual increases in U.S. total water withdrawals since the USGS began reporting in 1950, water use peaked in 1980, declined through 1990, and has remained relatively stable since then. (See Table 2.9, Figure 2.6, and Figure 2.7.) From 1995 to 2000 (the latest data available), a period that experienced a 7% increase in U.S. population, total off-stream water use increased only 2%. Water use for public supply increased by 8%, irrigation by 2%, and thermoelectric power use by 3%.

Hutson et al. state that the general increase in water use from 1950 to 1980 and the decrease from 1980 to 2000 can be attributed to several factors, including:

- Expansion of irrigation systems and increases in energy development from 1950 to 1980.

- The development and increasing use of two irrigation methods—center-pivot irrigation systems and drip

TABLE 2.9

Water use trends, selected years 1950–2000

[In billion gallons per day (thousand million gallons per day); rounded to two significant figures for 1950–80, and to three significant figures for 1985–2000; percentage change is calculated from unrounded number]

	Year											Percentage change
	1950[a]	1955[b]	1960[c]	1965[d]	1970[d]	1975[c]	1980[c]	1985[c]	1990[c]	1995[c]	2000[c]	1995–2000
Population, in millions	150.7	164.0	179.3	193.8	205.9	216.4	229.6	242.4	252.3	267.1	285.3	+7
Offstream use:												
Total withdrawals	**180**	**240**	**270**	**310**	**370**	**420**	**440**	**399**	**408**	**402**	**408**	**+2**
Public supply	14	17	21	24	27	29	34	36.5	38.5	40.2	43.2	+8
Rural domestic and livestock:												
Self-supplied domestic	2.1	2.1	2.0	2.3	2.6	2.8	3.4	3.32	3.39	3.39	3.59	+6
Livestock and aquaculture	1.5	1.5	1.6	1.7	1.9	2.1	2.2	4.47[e]	4.50	5.49	[f]	—
Irrigation	89	110	110	120	130	140	150	137	137	134	137	+2
Industrial:												
Thermoelectric-power use	40	72	100	130	170	200	210	187	195	190	195	+3
Other industrial use	37	39	38	46	47	45	45	30.5	29.9	29.1	[g]	—
Source of water:												
Ground:												
Fresh	34	47	50	60	68	82	83	73.2	79.4	76.4	83.3	+9
Saline	[h]	.6	.4	.5	1.0	1.0	.9	.65	1.22	1.11	1.26	+14
Surface:												
Fresh	140	180	190	210	250	260	290	265	259	264	262	−1
Saline	10	18	31	43	53	69	71	59.6	68.2	59.7	61.0	+2

[a]48 states and District of Columbia, and Hawaii.
[b]48 states and District of Columbia.
[c]50 states and District of Columbia, Puerto Rico, and U.S. Virgin Islands.
[d]50 states and District of Columbia, and Puerto Rico.
[e]From 1985 to present this category includes water use for fish farms.
[f]Data not available for all states; partial total was 5.46.
[g]Commercial use not available; industrial and mining use totaled 23.2.
[h]Data not available.

SOURCE: Susan S. Hutson et al., "Table 14. Trends in Estimated Water Use in the United States, 1950–2000," in *Estimated Use of Water in the United States in 2000*, U.S. Department of the Interior, U.S. Geological Survey, 2004, http://pubs.usgs.gov/circ/2004//circ1268/pdf/circular1268.pdf (accessed January 2, 2007) and revision data February 7, 2005, http://pubs.usgs.gov/circ/2004//circ1268/control/revisions.html (accessed January 2, 2007)

irrigation (the application of water directly to the roots of plants)—that are more efficient in delivering water to crops than the traditional sprayer arms that project the water into the air, where much is lost to wind and evaporation.

- Higher energy prices in the 1970s and a decrease in groundwater levels in some areas increased the cost of irrigation water.

- A downturn in the farm economy in the 1980s, which reduced demands for irrigation water.

- New industrial technologies requiring less water, improved efficiency, increased water recycling, higher energy prices, and changes in the law to reduce pollution.

- Active conservation programs and increased awareness by the general public of the need to conserve water.

WATER USE—THE FUTURE

The usefulness and availability of water can fluctuate dramatically in natural systems. Both the quality and quantity of water resources need to be protected for the nation's present and future generations. Furthermore, even though current water use can be determined, total water needs for most uses are changing. Water use is dependent on prices, technology, customs, and regulations. As such, water use data are good indicators of where and how the nation consumes water, but they are not necessarily good predictors of future water use trends.

Although the United States is not running out of water, the era of free and easily developed water supplies has ended for much of the country; in some areas water use is approaching or has exceeded the available supply.

Projections of freshwater usage by use-category are presented by Thomas C. Brown in *Past and Future Freshwater Use in the United States* (December 1999, http://www.fs.fed.us/rm/pubs/rmrs_gtr039.pdf). According to Brown, by 2040 freshwater usage in the United States will reach 364 Bgal/d. This figure represents a 7% increase over the 1995 usage rate of 340 Bgal/d.

Brown's projections anticipate increased freshwater usage for livestock and domestic and public water services. (See Figure 2.8.) Water usage for thermoelectric generation

FIGURE 2.6

Population and water use trends by source, selected years 1950–2000

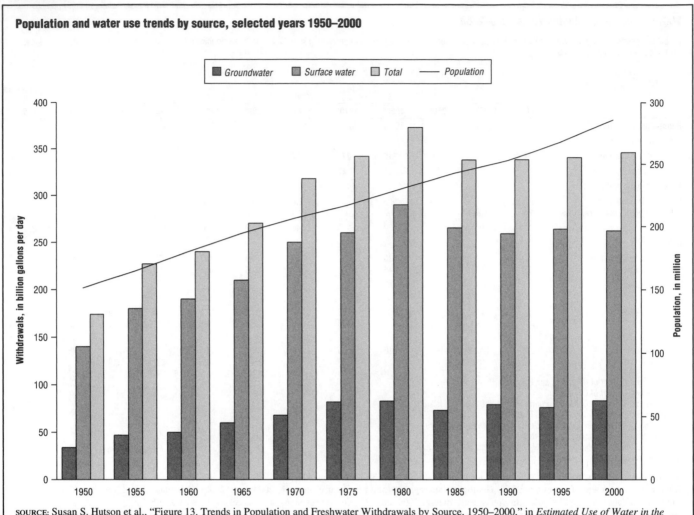

SOURCE: Susan S. Hutson et al., "Figure 13. Trends in Population and Freshwater Withdrawals by Source, 1950–2000," in *Estimated Use of Water in the United States in 2000,* U.S. Department of the Interior, U.S. Geological Survey, 2004, http://pubs.usgs.gov/circ/2004//circ1268/pdf/circular1268.pdf (accessed January 2, 2007) and revision data February 7, 2005, http://pubs.usgs.gov/circ/2004//circ1268/control/revisions.html (accessed January 2, 2007)

will rise slightly. The dark bars in Figure 2.8 indicate past withdrawals and the light bars indicate future withdrawals (projected as of 1999). The dots show levels of related factors, with dark dots showing past levels and light dots future levels. The related factor and its scale is on the right.

Freshwater usage for livestock and domestic and public water services is projected to increase at about the same rate as the U.S. population. Water usage for thermoelectric generation will rise only slightly, even though the kilowatt hours generated will increase much more substantially. In 1995 freshwater withdrawals for thermoelectric use were 132 Bgal/d; in 2040 the withdrawals are estimated at 142 Bgal/d. Although water withdrawals per day are projected to be higher in 2040 than they were in 1995, the increased kilowatt hours produced will result in a decreasing water withdrawal per kilowatt hour of electricity that will be produced.

The quantity of water used by industry and for commercial applications is anticipated to remain stable

through 2040. (See Figure 2.8.) David G. Lenze and Kathy Albetski report in "State Personal Income 2006" (March 27, 2007, http://www.bea.gov/newsreleases/regional/spi/2007/pdf/spi0307.pdf) that per capita income grew 5.2% in 2006, up from 4.2% in 2005. With continued annual growth in per capita income and with industrial and commercial withdrawals remaining stable, the result will be a decrease in water withdrawals for industrial and commercial uses per dollar of income.

Through 2040 the amount of irrigated acreage is expected to increase modestly, from 57.9 million acres in 1995 to 62.4 million acres in 2040. (See Figure 2.8.) Water withdrawals, however, are expected to decline modestly, from 134 Bgal/d in 1995 to 130 Bgal/d in 2040.

Overall, per capita freshwater withdrawals are projected to decline through 2040, although the net change will be an increase of about 7%. (See Figure 2.8.) Several factors are expected to contribute to the lower per capita freshwater usage rates. According to Brown, the two most

FIGURE 2.7

Water use trends by usage category, selected years 1950–2000

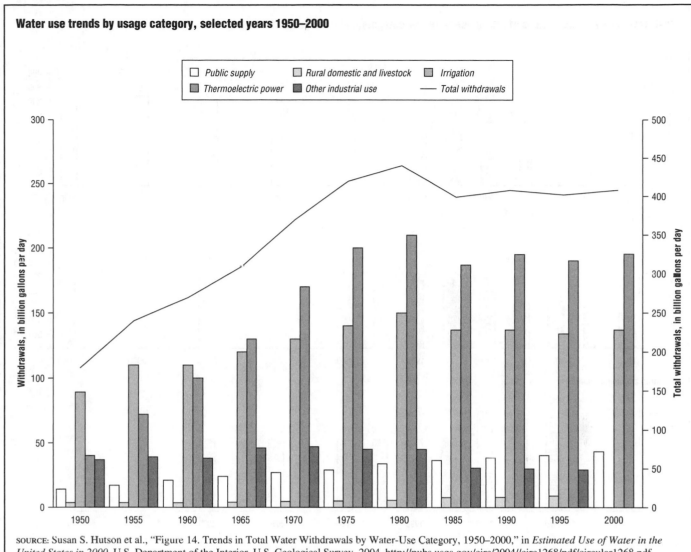

SOURCE: Susan S. Hutson et al., "Figure 14. Trends in Total Water Withdrawals by Water-Use Category, 1950–2000," in *Estimated Use of Water in the United States in 2000*, U.S. Department of the Interior, U.S. Geological Survey, 2004, http://pubs.usgs.gov/circ/2004//circ1268/pdf/circular1268.pdf (accessed January 2, 2007) and revision data February 7, 2005, http://pubs.usgs.gov/circ/2004//circ1268/control/revisions.html (accessed January 2, 2007)

prominent factors are improved efficiencies projected for the municipal, industrial, and thermoelectric generating sectors, and reduced irrigation withdrawals.

Increasing awareness among the traditional users of water and the general public of the finite nature of clean water supplies, particularly freshwater, has resulted in growing conservation efforts and innovative approaches to water conservation and reclamation. Water conservation is the careful use and protection of water resources. Water reclamation, also called water recycling, is the treatment of wastewater so that it can be reused for certain purposes, such as landscape irrigation.

Water Conservation

Water conservation and reclamation efforts take place all over the United States. The following are a few examples to show the types of activities that have been undertaken and their outcomes.

In "How We Can Do It" (*Scientific American*, February 2001), Diane Martindale and Peter H. Gleick describe a massive water conservation project undertaken in New York City. To prevent a pending water crisis in the early 1990s, New York City needed an extra ninety million gallons of water per day, about 7% of the city's total daily use. Faced with the need to raise $1 billion for a new pump station to bring additional water from the Hudson River, the city came up with a cheaper alternative: reduce the demand on the current supply. Using a three-year toilet rebate program, budgeted at $295 million for up to 1.5 million rebates, the city sought to replace about one-third of the existing toilets that used five gallons per flush with the water-saving models that did the same job with less than two gallons. By the end of the program in 1997, 1.3 million inefficient toilets in 110,000 buildings had been replaced with low-flow toilets. The result was about a 29% reduction in water use per building per year. The low-flow toilets were estimated to save about seventy to ninety Mgal/d.

FIGURE 2.8

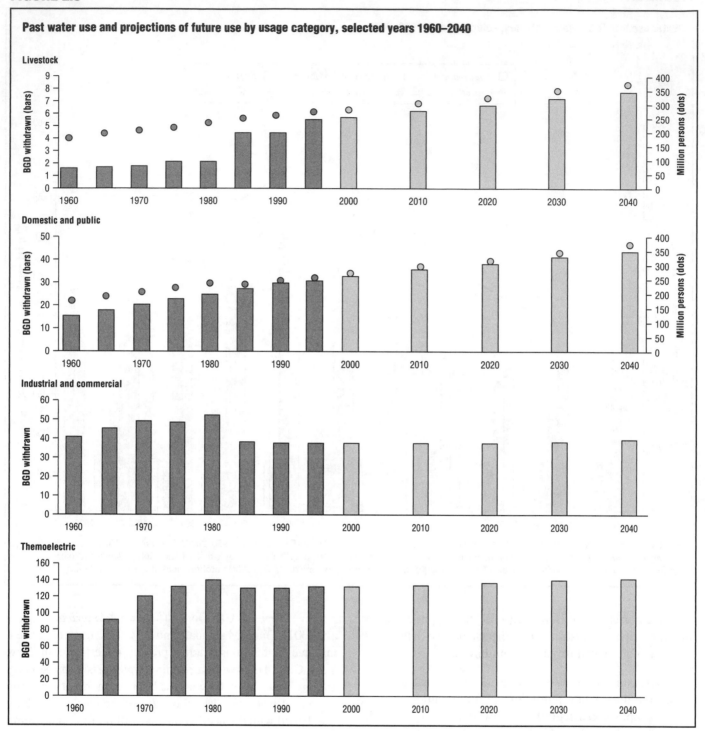

Past water use and projections of future use by usage category, selected years 1960–2040

Livestock

Domestic and public

Industrial and commercial

Themoelectric

New York City saved thirty to fifty Mgal/d of water from its leak detection program, two hundred Mgal/d from meter installation, and four Mgal/d from home inspections.

Concurrently, New York City had a water audit program under which property owners who wanted to reduce water use to keep bills down could request a free water efficiency survey from the company that oversaw the city's audit program. Inspectors checked for leaky plumbing, offered advice on retrofitting with water-efficient fixtures,

and distributed low-flow showerheads and water-efficient faucet aerators. Low-flow showerheads use about half the water of the old units. Faucet aerators, which replace the screen in the faucet head and add air to the spray, can reduce the flow from four gallons per minute to one gallon per minute. The company made hundreds of thousands of these inspections, saving an estimated eleven Mgal/d of water.

Overall, the program shows that water conservation works. Per person water use in New York City dropped

FIGURE 2.8

Past water use and projections of future use by usage category, selected years 1960–2040 [CONTINUED]

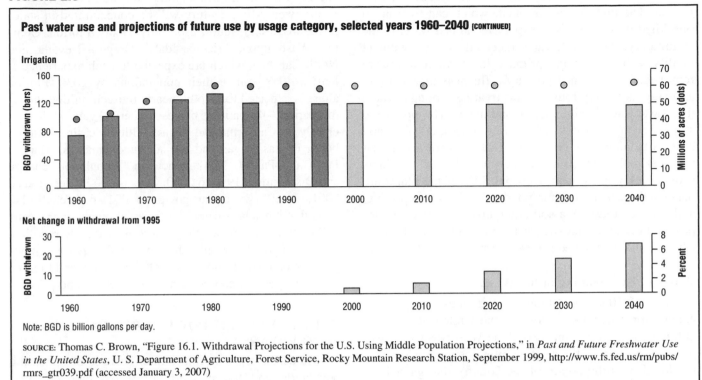

Note: BGD is billion gallons per day.

SOURCE: Thomas C. Brown, "Figure 16.1. Withdrawal Projections for the U.S. Using Middle Population Projections," in *Past and Future Freshwater Use in the United States*, U. S. Department of Agriculture, Forest Service, Rocky Mountain Research Station, September 1999, http://www.fs.fed.us/rm/pubs/rmrs_gtr039.pdf (accessed January 3, 2007)

from 195 to 169 gallons per person per day between 1991 and 1999, even though the city's population continued to grow.

In *Cases in Water Conservation: How Efficiency Programs Help Water Utilities Save Water and Avoid Costs* (July 2002, http://www.epa.gov/watersense/docs/utilityconservation_508.pdf), the EPA reports on conservation methods and incentives adopted by New York City and sixteen other North American locations. Each location was beset by problems such as a strain on the water supply, unaccounted-for water loss, and water shortages. One of the more significant conservation efforts took place in Gallitzin, Pennsylvania. In the mid-1990s the Gallitzin Water Authority reported water losses exceeding 70%. After identifying the major problems (high water loss, recurring leaks, high overall operational costs, low-pressure complaints, and unstable water entering the distribution system), the water authority developed accurate water production and distribution records using seven-day meter readings at its water plant and pump systems. Then it developed a system map to locate leaks. A leak detector located 95% of leaks in the water system. After repairs were made there was an 87% drop in unaccounted-for water loss and savings of $5,000 on total chemical costs and $20,000 on total annual power costs from 1994 to 1998.

Water Reclamation

The water department in Tampa, Florida, has been working to maximize the yield from its water supply. The South Tampa Area Reclaimed water project featured the use of high-quality reclaimed water from the Howard F. Curran Wastewater Treatment Facility to satisfy the demands of high-volume irrigation users in South Tampa. Water would be made available through a water system that would be separate from the drinking water supply to prevent any possibility of cross contamination. The project began as a grassroots effort by Westshore residents concerned about future water supplies. Important conditions of the project were voluntary participation in it, only citizens who wanted reclaimed water would have to participate in the project, and user fees would make the project self-supporting.

In the first four months of the project sign-up, forty-five hundred homeowners and businesses enrolled. Construction of Phase I began in 2002 and was near completion in 2005 at a cost of $28 million. The first users began drawing water from the system in July 2004. Recommended uses for reclaimed water are crop irrigation, lawn and landscape watering, washing cars, and general cleaning. In "Fast Facts" (2007, http://www.tampagov.net/dept_water/starproject/Fast_Facts.asp), the water department estimates that when fully operational, Phase I would save more than 3.2 Mgal/d of potable water each day during the dry season, and additional phases are planned to extend the project to a wider geographical area.

The San Diego County Water Authority (2006, http://www.sdcwa.org/manage/recycled-facilities.phtml) operates the San Pasqual water reclamation plant, which is located in the San Pasqual Valley near Escondido, California. The purpose of the plant, which can treat up to one Mgal/d of water, is to supply reclaimed water to the community. The wastewater received at the treatment facility is treated to the primary level when solids are removed. The screened primary effluent is then fed into as many as twenty-four aquatic treatment ponds, where the wastewater is biologically stabilized. The ponds are stocked with water hyacinth, mosquito fish, crayfish, and other organisms to create an aquatic ecosystem that removes pollutants from wastewater. Water hyacinths grow quickly in the ponds. About 50% of the plants are harvested weekly from the ponds, dried for composting, and sold for reuse as a soil amendment. After the water passes through the aquatic treatment ponds, it is clarified, filtered, and disinfected for use in irrigation and research.

Water Crisis Looming in the West?

Table 2.10 shows cumulative estimates of population change for states and regions of the United States from 2000 to 2006. During this period the West was the fastest-growing portion of the country by percentage. However, the South was the region of the country that gained the most people. Thus, both the West and the South vie for first and second place as the fastest-growing regions of the United States.

Percentage-wise, Nevada (24.9%) was the state that grew the most from 2000 to 2006, followed by Arizona (20.2%), Georgia (14.4%), Utah (14.2%), Idaho (13.3%), Florida (13.2%), Texas (12.7%), Colorado (10.5%), North Carolina (10.1%), and Delaware (8.9%), respectively. (See Table 2.10.) The state that grew the most in numbers of new residents was Texas (2.6 million), followed by California (2.5 million), Florida (2.1 million), Georgia (1.1 million), Arizona (1 million), North Carolina (810,000), Virginia (564,000), Washington (501,000), Nevada (497,000), and Colorado (451,000), respectively. Louisiana lost population during this period, with much of that decline attributed to Hurricane Katrina and its aftermath.

According to the U.S. Bureau of the Census, in "Louisiana Loses Population; Arizona Edges Nevada as Fastest-Growing State" (December 22, 2006, http://www.census.gov/Press-Release/www/releases/archives/population/007910.html), Arizona was the fastest-growing state with its population growing 3.6% in 2006. Nevada was a close second with its population growing 3.5%. The West was the fastest-growing region that year; its population grew by 1.5%. The South was the next fastest-growing region with a population increase of 1.4% In contrast, the population increase in the Midwest was 0.4% and in the Northeast was 0.1%

In *Interim State Population Projections, 2005* (April 21, 2005, http://www.census.gov/population/projections/PressTab7), Census Bureau projections show that population growth in the United States for the first thirty years of the twenty-first century will be concentrated in states in the West and South, especially Texas, Florida, Georgia, Washington, Arizona, Idaho, Oregon, Nevada, and North Carolina, which are expected to each experience at least a 40% gain in their populations by 2030. Florida alone is expected to leap by nearly thirteen million during this period—the nation's biggest numerical gain—followed closely by California and Texas. With the casino building boom in Las Vegas and the surrounding areas, Marc J. Perry and Paul J. Mackun note in "Population Change and Distribution, 1990 to 2000: Census 2000 Brief" (April 2001, http://www.census.gov/prod/2001 pubs/c2kbr01-2 .pdf) that Nevada became the fastest-growing state in the 1990s and Las Vegas the fastest-growing metropolitan area. This population growth is expected to put enormous pressure on natural resources, including water, and to force huge changes in water consumption practices and prices.

WHERE WATER IS POWER—INTERNATIONAL WATER WARS?

As usable water becomes rarer because of increasing population and the pollution of water supplies, it is expected to become a commodity such as iron or oil. Peter Allison, in "World Overview: Water Wars—The Global Viewpoint" (2006, http://www.itt.com/waterbook/world .asp), states that "over 20 countries depend on the flow of water from other nations for much of their supply. And more than 300 of the world's river basins are shared by two or more countries." All are potential objects of world political power struggles over this critical resource.

Africa, the Middle East, and South Asia are three areas of the world that are particularly short of water. Other dry areas include the southwest of North America, limited areas in South America, and large parts of Australia.

The rivers in the Middle East are the lifeblood of an arid region. Freshwater has never come easily to this area. Rainfall occurs only in winter and drains quickly through the parched land. Most Middle Eastern countries are joined by common aquifers. The United Nations cautions that future wars in the Middle East can be fought over water.

Since April 2001 tensions between Israel and Lebanon have been escalating over the Lebanese construction of a pumping station along the Hasbani River. The Hasbani flows into the Sea of Galilee, which is Israel's primary freshwater reservoir. In 2001 Israel was undergoing a water crisis, and the sea was at its lowest level ever. Tensions periodically arise as droughts tighten supply in the area, but since mid-2005 the tensions between Israel and Lebanon regarding Lebanon's pumping of water from the Hasbani River have been handled diplomatically, with the assistance of the European Union and the United Nations.

TABLE 2.10

Estimates of population change by region, state, and Puerto Rico, 2000–06

Geographic area	Population estimates		Change, 2000 to 2006		National ranking of states			
					Population estimates		Change, 2000 to 2006	
	July 1, 2006	April 1, 2000 estimates base	Number	Percent	July 1, 2006	April 1, 2000 estimates base	Number	Percent
United States	299,398,484	281,424,602	17,973,882	6.4	(X)	(X)	(X)	(X)
Northeast	54,741,353	53,594,784	1,146,569	2.1	4	4	4	4
Midwest	66,217,736	64,395,194	1,822,542	2.8	3	2	3	3
South	109,083,752	100,235,846	8,847,906	8.8	1	1	1	2
West	69,355,643	63,198,778	6,156,865	9.7	2	3	2	1
Alabama	4,599,030	4,447,351	151,679	3.4	23	23	27	34
Alaska	670,053	626,931	43,122	6.9	47	48	42	17
Arizona	6,166,318	5,130,632	1,035,686	20.2	16	20	5	2
Arkansas	2,810,872	2,673,398	137,474	5.1	32	33	28	22
California	36,457,549	33,871,653	2,585,896	7.6	1	1	2	15
Colorado	4,753,377	4,302,015	451,362	10.5	22	24	10	8
Connecticut	3,504,809	3,405,602	99,207	2.9	29	29	32	37
Delaware	853,476	783,600	69,876	8.9	45	45	37	10
District of Columbia	581,530	572,059	9,471	1.7	50	50	49	44
Florida	18,089,888	15,982,824	2,107,064	13.2	4	4	3	6
Georgia	9,363,941	8,186,816	1,177,125	14.4	9	10	4	3
Hawaii	1,285,498	1,211,537	73,961	6.1	42	42	36	20
Idaho	1,466,465	1,293,956	172,509	13.3	39	39	23	5
Illinois	12,831,970	12,419,647	412,323	3.3	5	5	11	36
Indiana	6,313,520	6,080,517	233,003	3.8	15	14	21	28
Iowa	2,982,085	2,926,382	55,703	1.9	30	30	40	41
Kansas	2,764,075	2,688,824	75,251	2.8	33	32	35	38
Kentucky	4,206,074	4,042,285	163,789	4.1	26	25	24	27
Louisiana	4,287,768	4,468,958	−181,190	−4.1	25	22	51	51
Maine	1,321,574	1,274,923	46,651	3.7	40	40	41	31
Maryland	5,615,727	5,296,506	319,221	6.0	19	19	14	21
Massachusetts	6,437,193	6,349,105	88,088	1.4	13	13	33	46
Michigan	10,095,643	9,938,480	157,163	1.6	8	8	26	45
Minnesota	5,167,101	4,919,492	247,609	5.0	21	21	19	23
Mississippi	2,910,540	2,844,656	65,884	2.3	31	31	38	40
Missouri	5,842,713	5,596,683	246,030	4.4	18	17	20	25
Montana	944,632	902,195	42,437	4.7	44	44	43	24
Nebraska	1,768,331	1,711,265	57,066	3.3	38	38	39	35
Nevada	2,495,529	1,998,257	497,272	24.9	35	35	9	1
New Hampshire	1,314,895	1,235,786	79,109	6.4	41	41	34	18
New Jersey	8,724,560	8,414,347	310,213	3.7	11	9	16	30
New Mexico	1,954,599	1,819,046	135,553	7.5	36	36	29	16
New York	19,306,183	18,976,821	329,362	1.7	3	3	13	43
North Carolina	8,856,505	8,046,491	810,014	10.1	10	11	6	9
North Dakota	635,867	642,200	−6,333	−1.0	48	47	50	50
Ohio	11,478,006	11,353,145	124,861	1.1	7	7	31	48
Oklahoma	3,579,212	3,450,654	128,558	3.7	28	27	30	29
Oregon	3,700,758	3,421,436	279,322	8.2	27	28	18	12
Pennsylvania	12,440,621	12,281,054	159,567	1.3	6	6	25	47
Rhode Island	1,067,610	1,048,319	19,291	1.8	43	43	46	42
South Carolina	4,321,249	4,011,816	309,433	7.7	24	26	17	14
South Dakota	781,919	754,844	27,075	3.6	46	46	44	33
Tennessee	6,038,803	5,689,262	349,541	6.1	17	16	12	19
Texas	23,507,783	20,851,790	2,655,993	12.7	2	2	1	7
Utah	2,550,063	2,233,198	316,865	14.2	34	34	15	4
Vermont	623,908	608,827	15,081	2.5	49	49	47	39
Virginia	7,642,884	7,079,030	563,854	8.0	12	12	7	13
Washington	6,395,798	5,894,140	501,658	8.5	14	15	8	11
West Virginia	1,818,470	1,808,350	10,120	0.6	37	37	48	49
Wisconsin	5,556,506	5,363,715	192,791	3.6	20	18	22	32
Wyoming	515,004	493,782	21,222	4.3	51	51	45	26
Puerto Rico	3,927,776	3,808,603	119,173	3.1	(X)	(X)	(X)	(X)

Note: The April 1, 2000 population estimates base reflects changes to the Census 2000 population from the count question resolution program and geographic program revisions. (X) Not applicable.

SOURCE: "Table 2. Cumulative Estimates of Population Change for the United States, Regions, States and Puerto Rico and Region and State Rankings: April 1, 2000 to July 1, 2006," U.S. Census Bureau, Population Division, December 22, 2006, http://www.census.gov/popest/states/tables/NST-EST2006-02.xls (accessed January 2, 2007)

The oil-rich Middle Eastern nation of Kuwait has little water, but it has the money to secure it. To use seawater, Kuwait has constructed six large-scale, oil-powered desalination plants. According to the U.S. Commercial Service, in "Water Resources Equipment (WRE)" (2007, http://www.buyusa.gov/saudiarabia/en/113.html), Saudi Arabia leads the world in desalination. In 2007 its thirty plants produced 30% of all the desalinated water in the world.

Water quality and water shortages are just two of the problems facing the world in the years to come. The September 11, 2001, terrorist attacks in the United States increased concerns about bioterrorism. On June 12, 2002, Congress passed the Public Health Security and Bioter-rorism Act of 2002. Title IV of the act (Drinking Water Security and Safety) mandates that every community water system that serves a population of more than thirty-three hundred people must:

- Conduct a vulnerability assessment.

- Certify and submit a copy of the assessment to the EPA administrator.

- Prepare or revise an emergency response plan that incorporates the results of the vulnerability assessment.

- Certify to the EPA administrator, within six months of completing the vulnerability assessment, that the system has completed or updated its emergency response plan.

CHAPTER 3
SURFACE WATER: RIVERS, STREAMS, AND LAKES

Most of the earth's water, about 97%, is the saltwater of the oceans. (See Figure 1.2 in Chapter 1.) By comparison, freshwater comprises only 3% of the earth's water, and surface water is only 0.3% of that 3%. Furthermore, rivers and lakes comprise 2% and 87%, respectively, of surface water. Thus, rivers contain only 0.00018% and lakes only 0.0078% of the earth's water. Nonetheless, this tiny fraction of the total water supply has shaped the course of human development. Throughout human history societies have depended on these surface water resources for food, drinking water, transportation, commerce, power, and recreation.

In 2000, out of a total of 345 billion gallons per day (Bgal/d) of the total freshwater consumption in the United States, water from streams, rivers, and lakes accounted for 262 Bgal/d (76%). (See Table 2.1 in Chapter 2.) The remaining 83.3 Bgal/d (24%) came from groundwater. Public utilities (public and private water suppliers) used 27.3 Bgal/d (63%) of surface water for their operations. (See Table 2.3 in Chapter 2.) Industries consumed 14.9 Bgal/d (76%) of surface freshwater for their requirements. (See Table 2.7 in Chapter 2.) Meanwhile, crop irrigation used 80 Bgal/d (58%) of surface water to water crops. (See Table 2.5 in Chapter 2.)

The withdrawal of surface water varies greatly throughout the United States. Figure 2.2 (see Chapter 2) shows that in 2000 California, Texas, and Florida used the most water per day, whereas the Dakotas used the least.

CHARACTERISTICS OF RIVERS AND LAKES
Rivers and Streams

The great rivers of the world have influenced human history. Settlements on rivers have thrived since earliest recorded history, with most of the world's great civilizations growing up along rivers. Flowing rivers provided water to drink, fish and shellfish to eat, dispersion and removal of wastes, and transport for goods. The bountiful supply of freshwater in flowing rivers is one of the primary reasons for the rapid growth of settlement, industry, and agriculture in the United States during both colonial and modern times.

Rivers and streams, unlike lakes, consist of flowing water. Perennial rivers and streams flow continuously, although the volume may vary with runoff conditions. Intermittent, or ephemeral, rivers and streams stop flowing for some period, usually because of dry conditions. Both large and small rivers and streams are an important part of the hydrologic cycle. (See Figure 1.3 in Chapter 1.)

Rivers receive water from rain and melting snow, from underground springs and aquifers, and from lakes. A large river is usually fed by tributaries (smaller rivers and streams), and so increases in size as it travels from its source, or origin. Its final destination may be an ocean, a lake, or sometimes open land, where the water simply evaporates. This phenomenon usually happens only with small rivers or streams.

As water flows down a river, it carries with it grains of soil, sand, and, where there is a strong current, small stones and other debris. These objects are important in two ways. First, as they are pulled along by the river's current, they grind against the bottom and sides of the riverbank and slowly cut the riverbed deeper and deeper into the earth, thereby changing the contour of the land. (The Grand Canyon is an example of how a river can carve the land.) Second, when the river reaches its destination (an ocean or lake), the flow is slowed and then stopped where the bodies of water meet, and the soil that has been carried along is deposited. These deposits are called sediment. Finely grained sediment is called silt.

Over long periods, sediment deposited at the mouths of rivers forms triangular-shaped areas called deltas. During flood conditions some of the sediment of the delta floodwaters is deposited on the low-lying land (floodplain) surrounding the river. Enriched with this sediment, the delta floodplain often provides a rich base for agriculture. The ancient civilizations of Egypt, for example, depended on the land surrounding the delta of the Nile River to grow their food supply. On the contrary, deposited sediment can become a nuisance by filling lakes and harbors and smothering aquatic life. Many ports and harbors in the United States must be dredged regularly to remove deposits that would otherwise obstruct navigation.

The U.S. Geological Survey reports in "Lengths of the Major Rivers" (August 28, 2006, http://ga.water.usgs.gov/edu/riversofworld.html) that the two longest rivers in the world are the Nile (4,132 miles) and the Amazon (4,000 miles). The Mississippi-Missouri river system is the third longest in the world. Taken separately, the Missouri River (2,540 miles) is the longest in the United States and the Mississippi (2,340 miles) the second longest. (See Table 3.1.)

The Mississippi River has an enormous watershed, the land from which it receives runoff water from rainfall or snowmelt. According to the Mississippi National River and Recreation Area (April 12, 2004, http://www.nps.gov/archive/miss/features/factoids/), this watershed is between 1.2 and 1.8 million square miles, or about 41% of the total land area of the lower forty-eight states. The Mississippi River is considered the largest river in the United States by the measure of the average volume at its mouth. (See Table 3.1.) The "Mighty Mississippi," as it is often called, discharges its water into the Gulf of Mexico at an average rate of 593,000 cubic feet per second.

Lakes

Unlike rivers, lakes are depressions in the earth that hold water for extended periods. Reservoirs are human-made lakes created when a dam is built on a river. They are generally used to store water for uses such as drinking, irrigation, or producing electricity. Often, reservoir areas are used for recreation as well. Some ponds are made for livestock watering, fire control, storm water management, duck and fish habitat, and recreation.

The source of the water in lakes, reservoirs, and ponds may be rivers, streams, groundwater, rainfall, melting snow runoff, or a combination of these. Any of these sources may carry contaminants. Because water exits from these water bodies at a slow rate, pollutants can become trapped and build up.

Many of the world's lake beds were formed during the Ice Age, when advancing and retreating glaciers gouged holes in the soft bedrock and spread dirt and debris in uneven patterns. Some lakes fill the craters of extinct volcanoes, and others have formed in the shallow basins of ocean bottoms uplifted by geological activity to become part of the earth's solid surface.

TABLE 3.1

Largest rivers in the United States in length and average discharge volume at mouth

River	Location at mouth	Average discharge at mouth (cfs)	Length[a] (mi.)
Mississippi	LA	593,000	2,340
St. Lawrence	Canada	348,000	1,900
Ohio	IL, KY	281,000	1,310
Columbia	WA	265,000	1,240
Yukon	AK	225,000	1,980
Missouri	MO	76,200	2,540
Tennessee	KY	68,000	886
Mobile	AL	67,200	774
Kuskoswim	AK	67,000	724
Copper	AK	59,000	286
Atchafalaya	LA	58,000	140[b]
Snake	WA	56,900	1,040
Stikine	AK	56,000	379
Red	LA	56,000	1,290
Susitna	AK	51,000	313
Tanana	AK	41,000	659
Arkansas	AR	41,000	1,460
Susquehanna	MD	38,200	447
Willamette	OR	37,400	309
Nushagak	AK	36,000	285

Note: cfs is cubic feet per second.
[a]Including headwaters and sections in Canada.
[b]Below Mississippi diversion, without headwaters.

SOURCE: J.C. Kammerer, "Largest Rivers in the United States in Discharge, Drainage Area, or Length," in Water Fact Sheet: Largest Rivers in the United States, U.S. Department of the Interior, U.S, Geological Survey, May 1990, http://pubs.usgs.gov/of/1987/ofr87-242/pdf/ofr87242.pdf (accessed January 4, 2007)

As soon as a lake or pond is formed, it is destined to die. "Death" occurs over a long time, particularly in the case of large lakes. Soil and debris carried by in-flowing rivers and streams slowly build up the basin floor. At the same time, water is removed by out-flowing rivers and streams, whose channels become ever wider and deeper, allowing them to carry more water away. Even lakes that have no river inlets or outlets eventually fill with soil eroded from the surrounding land.

FRESHWATER VERSUS SALTWATER. According to the U.S. Environmental Protection Agency (EPA), in The Hydrologic (Water) Cycle (2007, http://www.epa.gov/OGWDW/kids/wsb/pdfs/9121.pdf), the large freshwater lakes of the world contain about 30,000 cubic miles of water and cover a combined surface area of about 330,000 square miles. In "NatureWorks: Lakes" (2007, http://www.nhptv.org/Natureworks/nwep7c.htm), New Hampshire Public Television indicates that by surface area Lake Superior, which is located on the U.S.-Canadian border, is the largest freshwater lake in the world. Lake Baikal in Asiatic Russia, however, is the deepest freshwater lake in the world, while Oregon's Crater Lake is the deepest in the United States.

Harvey A. Bootsma and Robert E. Hecky report in "A Comparative Introduction to the Biology and Limnology of the African Great Lakes" (Journal of Great Lakes

Research, 2003) that the large lakes of Africa (Victoria, Tanganyika, and Malawi) contain about 32% of the volume of all freshwater lakes on Earth (7,062 cubic miles [mi^3] or 29,435 cubic kilometers [km^3] of 21,832 mi^3 or 91,000 km^3). The North American Great Lakes (Superior, Michigan, Huron, Erie, and Ontario) hold slightly less, or about 25% of the volume of the freshwater lakes (5,472 mi^3 or 22,807 km^3). Lake Baikal holds slightly more water (5,662 mi^3 or 23,600 km^3) than the North American Great Lakes, holding about 26% of the volume of Earth's freshwater lakes. Together, these nine lakes hold 83% of the volume of all freshwater lakes.

The Geological Survey notes in *Where Is Earth's Water Located?* (August 28, 2006, http://ga.water.usgs .gov/edu/earthwherewater.html) that the saline (saltwater) lakes of the world contain almost as much water as the freshwater lakes (20,488 mi^3 or 85,400 km^3). Rene P. Schwarzenbach, Philip M. Gschwend, and Dieter M. Imboden, in "Ponds, Lakes, and Oceans" (*Environmental Organic Chemistry*, 2003), indicate that of that volume, however, nearly 92% is in the Caspian Sea (18,761 mi^3 or 78,200 km^3), which borders Russia and Iran. Most of the remainder is in lakes in Asia. North America's shallow Great Salt Lake is comparatively insignificant; it varies in depth and volume with climatic conditions, so exact comparisons are difficult.

NEED FOR POLLUTION CONTROL

People have always congregated on the shores of lakes and rivers. Benefiting from the many advantages of nearby water sources, they established permanent homes, then towns, cities, and industries. One of these advantages has been that lakes or rivers were convenient places to dispose of waste. As industrial societies developed, the amount of waste became enormous. Frequently, the wastes contained synthetic and toxic materials that could not be assimilated by the waters' ecosystems. Millions of tons of sewage, pesticides, chemicals, and garbage were dumped into waterways worldwide until there were few that were not contaminated to some extent. Some were—and some still are—contaminated to the point of ecological death—that is, that they are unable to sustain a balanced aquatic-life system.

Clean Water Act

On October 18, 2002, President George W. Bush proclaimed the beginning of the Year of Clean Water in commemoration of the thirtieth anniversary of the signing of the Clean Water Act (CWA), the full name of which is the Federal Water Pollution Control Act.

The CWA was enacted by Congress in 1972 in response to growing public concern over the nation's polluted waters. Although the Federal Water Pollution Control Act was enacted originally in 1948, it was amended many times and then reorganized and expanded in 1972. It continues to be amended almost every year.

The original 1948 legislation was intended to eliminate or reduce the pollution of interstate waters and improve the sanitary condition of surface and underground waters. However, on June 22, 1969, the Cuyahoga River in Cleveland, Ohio, burst into flames, the result of oil and debris that had accumulated on the river's surface. Clearly, the legislation was not achieving its desired goal.

The objective of the CWA was to "restore and maintain the chemical, physical, and biological integrity of the Nation's waters." Primary authority for the enforcement of this law lies with the EPA.

The CWA requires that, where attainable, water quality be such that it "provides for the protection and propagation of fish, shellfish, and wildlife and provides for recreation in and on the water." This requirement is referred to as the act's "fishable/swimmable" goal. Many people credit the CWA with reversing, in a single generation, what had been a decline in the health of the nation's water since the mid-nineteenth century.

Assessing and Monitoring the Quality of Water

Under section 305(b) of the CWA, the states are required to submit assessments of their water quality to the EPA every two years. The EPA is required to summarize this information in a biennial report to Congress. The EPA met this mandate by researching and compiling the *National Water Quality Inventory Report to Congress* every two years. Along with reporting to Congress, the purpose of the biennial reports was to inform the public about water quality in the United States. The *National Water Quality Inventory* reports characterized water quality, identified water quality problems of national significance, and described programs implemented to restore and protect the nation's water.

In the 2000 report *Water Quality: Key EPA and State Decisions Limited by Inconsistent an Incomplete Data* (March 2000, http://www.gao.gov/new.items/rc00054.pdf), the U.S. General Accounting Office (GAO; now the U.S. Government Accountability Office) stated that "the *National Water Quality Inventory* does not accurately portray water quality conditions nationwide." The GAO noted that states collectively assess only a small percentage of waters and that monitoring assessments and the interpretation of assessment results varied across states. Thus, the GAO indicated that "the information in the *Inventory* cannot be meaningfully compared across states." In addition, the data collected by the states were often insufficient to enable them to pinpoint and clean up their water quality problems.

As a result, the EPA, states, tribes, and other federal agencies began collaborating on a new process to monitor the nation's waterways. Following the publication of the *2000 National Water Quality Inventory* (August 2002, http://www.epa.gov/305b/2000report/), the EPA entered a transition period in the gathering and analysis of water quality data in nationally consistent, statistically valid assessment reports. Its new reporting schedule is discussed in "Schedule for Statistically Valid Surveys of the Nation's Waters" (December 5, 2005, http://www.epa .gov/owow/monitoring/guide.pdf). As of early 2007 the only new reports available were *The National Coastal Condition Report II (2005)* (December 2004, http:// www.epa.gov/owow/oceans/nccr/2005/downloads.html) and *The Wadeable Streams Assessment: A Collaborative Survey of the Nation's Streams* (December 2006, http://www.epa .gov/owow/streamsurvey/WSA_Assessment_Dec2006.pdf). The *Wadeable Streams Assessment* serves as a basis for most of the data in this chapter, and the *National Coastal Condition Report II* is used in Chapter 6.

In "An Introduction to Water Quality Monitoring" (March 21, 2007, http://www.epa.gov/owow/monitoring/ monintr.html), the EPA reports that the five major purposes for water quality assessment and monitoring were to:

- characterize waters and identify changes or trends in water quality over time;

- identify specific existing or emerging water quality problems;

- gather information to design specific pollution prevention or remediation programs;

- determine whether program goals—such as compliance with pollution regulations or implementation of effective pollution control actions—are being met; and

- respond to emergencies, such as spills and floods.

Agriculture Takes up the Challenge

According to the *2002 Census of Agriculture* (March 20, 2006, http://www.nass.usda.gov/Census_of_Agriculture/ index.asp) by the U.S. Department of Agriculture (USDA), about 938 million acres, or roughly half of the continental United States, is used for agricultural production. Cropland accounts for 46% of the acreage, and pasture and range land make up another 40%. Agricultural land use is recognized in many jurisdictions and localities throughout the United States as the most desirable land use for economic, environmental, and social reasons. At the same time, the public and the agricultural community recognize that agricultural practices are a source of nonpoint pollution nationwide. (Nonpoint source pollutants enter bodies of water over large areas rather than at single points. Nonpoint sources of water pollution include agricultural runoff and soil erosion.) This situation presents a challenge to water quality management efforts.

The agricultural community shares in the growing national concern over water quality degradation. There

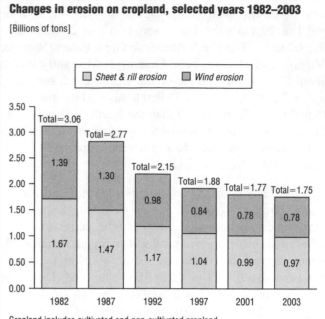

FIGURE 3.1

Changes in erosion on cropland, selected years 1982–2003

[Billions of tons]

☐ *Sheet & rill erosion* ■ *Wind erosion*

Total=3.06 (1982): Wind 1.39, Sheet & rill 1.67
Total=2.77 (1987): Wind 1.30, Sheet & rill 1.47
Total=2.15 (1992): Wind 0.98, Sheet & rill 1.17
Total=1.88 (1997): Wind 0.84, Sheet & rill 1.04
Total=1.77 (2001): Wind 0.78, Sheet & rill 0.99
Total=1.75 (2003): Wind 0.78, Sheet & rill 0.97

Cropland includes cultivated and non-cultivated cropland

SOURCE: "Erosion on Cropland by Year," in *National Resources Inventory 2003 Annual NRI*, U.S. Department of Agriculture, Natural Resources Conservation Service, May 2006, http://www.nrcs.usda.gov/ TECHNICAL/land/nri03/SoilErosion-mrb.pdf (accessed January 2, 2007). Surveys were done in cooperation with Iowa State University's Center for Survey Statistics and Methodology.

has been a steady increase in the use of best management practices and implementation of farm water quality plans to protect wetlands and water bodies. The success of this effort can be seen in the decrease in soil erosion of U.S. cropland between 1982 and 2003. (See Figure 3.1.) In 1982 a total of 3.1 billion tons of cropland eroded from the nation's agricultural areas. By 2003 this figure had dropped to 1.7 billion tons, a 43% decrease. The USDA, with state and local agencies, is providing technical assistance and financial incentives through many programs to help farmers balance good stewardship of natural resources with market demands. Technical assistance through these programs has had success in getting farmers to voluntarily adopt more environmentally sensitive practices.

ASSESSING WATER QUALITY OF WADEABLE STREAMS

What Are Wadeable Streams?

Wadeable streams are exactly what their name suggests: flowing bodies of water in which people can walk throughout. Researchers can perform tests and take samples of water in wadeable streams without a boat. The EPA notes in the *Wadeable Streams Assessment* that approximately 90% of stream and river miles in the United States are wadeable streams. Thus, an assessment of the water quality of the nation's wadeable streams is an excellent indicator of the water quality of much of the flowing freshwater in the country.

The *Wadeable Streams Assessment* covers streams within the lower forty-eight states. Wadeable stream assessments in Alaska, Hawaii, Puerto Rico, and Guam are not included in the 2006 report, but will be included in future reports. The EPA's assessment reports on the water quality of wadeable streams nationally and by dividing the nation into three regions: the West, the Plains and Lowlands, and the Eastern Highlands. (See Figure 3.2.) The West includes mountainous and dry regions; the Eastern Highlands are the mountainous regions east of the Mississippi River; and the Plains and Lowlands include low-elevation areas of the East and Southeast and the plains areas (vast grassland regions). The sampling areas were selected using techniques that provided a random sample having the full range of variation of the wadeable streams of the United States.

Indicators of the Biological Health of Freshwater Streams

To determine the biological health of freshwater streams, EPA researchers examined the biological condition of the aquatic macroinvertebrates living there. Aquatic macroinvertebrates are animals without backbones that live in water and can be seen with the naked eye, such as certain fly larvae, worms, and beetles. The number and types of the aquatic macroinvertebrates living in a stream reflect the biological condition of the water, because the organisms are exposed to the pollutants and various other stressors in the water. Certain species of macroinvertebrates can survive only in freshwater of good quality, whereas others can survive in good-, fair-, or poor-quality water. Fish live on the macroinvertebrates in a stream, so the presence of certain species of fish reflects not only the conditions in which certain fish species survive but also the presence of certain types of aquatic macroinvertebrates on which they typically feed.

In a simplified example, if researchers find a thriving community of mayfly larvae, riffle beetles, and trout, then the water quality is likely to be good. If these organisms are absent and crayfish, dragonfly nymph, and clams predominate, then the water quality is likely fair. An abundance of aquatic worms, leeches, black fly larvae, and catfish signal poor water quality. However, organisms that survive in fair- or poor-quality water can also survive in good-quality water, so determining not only the presence but also the relative abundance of species of organisms inhabiting the stream sample area is important.

Factors Responsible for Diminished Water Quality

For the 2006 *Wadeable Streams Assessment*, EPA researchers also measured factors responsible for diminished water quality, which are called aquatic indicators of

stress, stressor indicators, or, simply, stressors. Some stressors are naturally occurring, and some are the result of human activity. According to the EPA:

> Most physical stressors are created when we modify the physical habitat of a stream or its watershed, such as through extensive urban or agricultural development, excessive upland or bank erosion, or loss of streamside trees and vegetation. Examples of chemical stressors include toxic compounds (e.g., heavy metals, pesticides), excess nutrients (e.g., nitrogen and phosphorus), or acidity from acidic deposition or mine drainage. Biological stressors are characteristics of the biota that can influence biological integrity, such as the proliferation of non-native or invasive species (either in the streams and rivers, or in the riparian areas adjacent to these water bodies).

Water of good quality has low levels of pollutants and other stressors, and a high level of dissolved oxygen. Fair water quality has a higher level of pollutants and other stressors than good-quality water, and a lower level of dissolved oxygen. Poor water quality has even higher levels of pollutants and other stressors, and even lower levels of dissolved oxygen.

WATER QUALITY OF THE NATION'S STREAMS

Figure 3.2 shows a summary of the condition of wadeable streams across the country in 2006. The streams in the West had the best water quality, with 45.1% in good condition, whereas 29% of the streams of the Plains and Lowlands were in good condition, and 18.2% of those in the Eastern Highlands were in good condition. Overall, only 28.2% of the streams in the coterminous (lower forty-eight) states of the United States were in good condition.

The EPA notes that the water quality of the streams of the Eastern Highlands was of the greatest concern. Not only were a mere 18.2% in good condition but also over half (51.8%) were in poor condition. This compares to 40% in poor condition in the Plains and Lowlands, and 27.4% in poor condition the West. Overall, 41.9% of the nation's streams were in poor condition in 2006.

Macroinvertebrate Index of Biotic Condition

The Macroinvertebrate Index of Biotic Condition is a statistical measure that provides a total score based on six characteristics of the macroinvertebrates found in a particular water sample. This total score is one indicator of water quality. Researchers determine which species of macroinvertebrates are present in a particular water sample, the proportion of each, the level of diversity of species within the sample (high-quality water has a high level of diversity), and the feeding habits, habitats, and pollution tolerance of the species present. Each of these characteristics is an indicator of water quality, and together they provide a snapshot of the macroinvertebrate

FIGURE 3.2

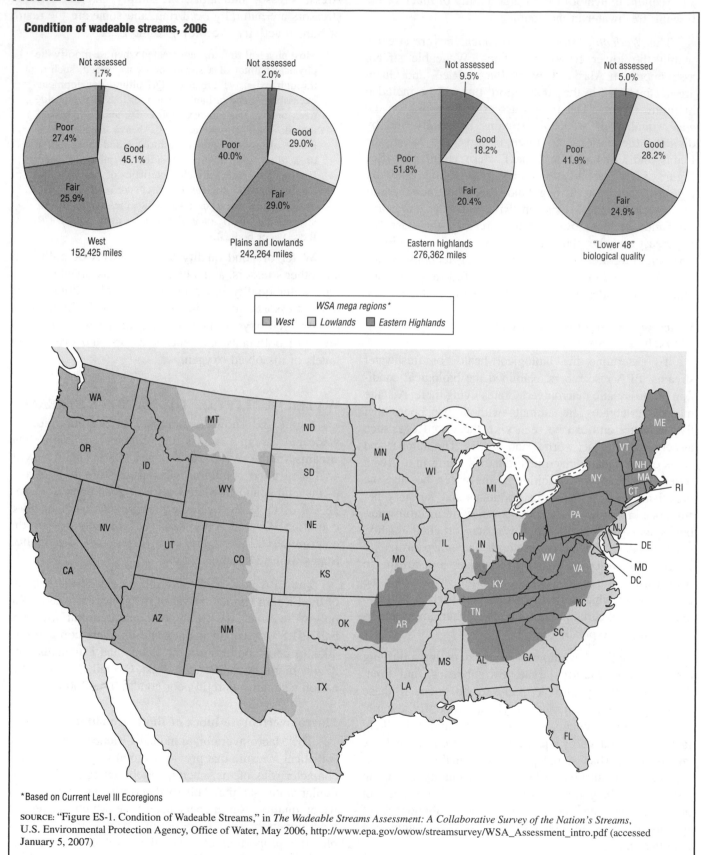

Condition of wadeable streams, 2006

Not assessed
1.7%

Poor
27.4%

Good
45.1%

Fair
25.9%

West
152,425 miles

Not assessed
2.0%

Poor
40.0%

Good
29.0%

Fair
29.0%

Plains and lowlands
242,264 miles

Not assessed
9.5%

Poor
51.8%

Good
18.2%

Fair
20.4%

Eastern highlands
276,362 miles

Not assessed
5.0%

Poor
41.9%

Good
28.2%

Fair
24.9%

"Lower 48"
biological quality

*WSA mega regions**
☐ West ☐ Lowlands ■ Eastern Highlands

*Based on Current Level III Ecoregions

source: "Figure ES-1. Condition of Wadeable Streams," in *The Wadeable Streams Assessment: A Collaborative Survey of the Nation's Streams*, U.S. Environmental Protection Agency, Office of Water, May 2006, http://www.epa.gov/owow/streamsurvey/WSA_Assessment_intro.pdf (accessed January 5, 2007)

"naturalness" in the portion of the stream tested. The researchers then use this total score, factor in the stream length represented by the study site, and use the data from all the study sites to compile the Macroinvertebrate Index ratings for the stream miles of a region and for the nation.

Figure 3.3 shows the Macroinvertebrate Index of Biotic Condition for the coterminous states as well as for the Eastern Highlands, the Plains and Lowlands, and the West. Nationally, 28.2% of the total miles of wadeable streams was in good condition in terms of macroinvertebrate life, 24.9% was in fair condition, and 41.9% was in poor condition. The Eastern Highlands had the most stream length in poor condition: 51.8%. The West had the most stream length in good condition: 45.1%.

Chemical Stressors

Four chemical stressors were assessed for the 2006 *Wadeable Streams Assessment*: phosphorus, nitrogen, salinity, and acidification. The levels of these stressors in the samples were compared with data from a set of "least disturbed" reference sites in each region to develop regional thresholds for all indicators.

PHOSPHORUS AND NITROGEN. Phosphorus and nitrogen are plant nutrients. When phosphorus and nitrogen enter bodies of water—usually as runoff from fertilized land, leaking septic systems, or sewage discharges—they promote aquatic plant and algal growth. Plant and algal growth can become excessive, a process called eutrophication. This excessive growth can result in waters clogged with plants and algae, which can look unsightly, slow water flow, and interfere with swimming and fishing. Mats of algae can grow on the surface of the water, blocking light to plants beneath. When these plants die, bacteria degrade them, using oxygen in the process and diminishing the concentration of oxygen in the water available for aquatic macroinvertebrates and fish. With lowered dissolved oxygen in the water, many fish and invertebrates die, worsening the situation.

Phosphorus is a common component of fertilizers and was routinely found in laundry detergents until the industry removed phosphates from its products in 1994. However, phosphates are still found in dishwashing detergents and in some cleaners. These phosphates wash down the drain during or after use and enter either septic systems or sewage treatment plants. From there, the phosphates can end up in bodies of water as they leach into the ground from septic systems or are discharged into streams with treated wastewater from sewage treatment plants. Agricultural runoff containing phosphate fertilizers is also a common source of added phosphates in bodies of water. The EPA summarizes in the *Wadeable Streams Assessment* by stating that "high phosphorus concentrations in streams may be associated with poor agricultural practices, urban runoff, or point-source discharges (e.g., effluents from sewage treatment plants)."

Figure 3.4 shows the percent of stream miles with low, medium, and high levels of phosphorus compared to regional references. "Low" means the concentrations were most similar to the reference (most natural) condition. "Medium" and "high" had statistical bases, but they can be thought of as above the regional reference (medium) and much above the regional reference (high). Nationally, nearly one-third (30.9%) of all stream miles had high levels of total phosphorus. However, nearly half (48.8%) of all stream miles had low levels of this chemical stressor.

Regionally, the highest percentage of stream miles with high levels of phosphorus was in the Eastern Highlands, where 42.6% of all stream miles had high levels of the chemical. (See Figure 3.4.) The Plains and Lowlands had 24.9% of its stream miles high in phosphorus, whereas the West had the least number of stream miles with high levels of this chemical stressor: 18.5%.

Nitrogen is another plant nutrient, as noted previously. It finds it way into streams primarily from agricultural runoff (it is found in fertilizer), wastewater and animal waste (it is a waste product from the digestion of protein), and atmospheric deposition (it is released into the air when fossil fuels, such as gasoline and coal, are burned). Nitrogen is particularly important as a contributor to the rapid, excessive growth of algae along coastal waters and in estuaries, where freshwater meets the saltwater of the ocean.

Figure 3.5 shows the percent of stream miles with low, medium, and high levels of nitrogen compared to regional references. The statistics are similar to those for phosphorus. Nationally, nearly one-third (31.8%) of all stream miles had high levels of total nitrogen. However, close to half (43.3%) of all stream miles had low levels of this chemical stressor.

Regionally, the highest percentage of stream miles with high levels of nitrogen was in the Eastern Highlands, where 42.4% of all stream miles had high levels of the chemical. (See Figure 3.5.) The Plains and Lowlands had about one-quarter (27.1%) of their stream miles high in nitrogen, whereas the West had the least number of stream miles with high levels of this chemical stressor: 20.5%.

SALINITY. Excessive salinity in freshwater streams generally occurs because water is lost from the stream, not because excessive salts enter the stream. This happens when the evaporation rate of stream water is high. Already high salinity from evaporation can be made higher by repeated water withdrawals for irrigation or other purposes.

Figure 3.6 shows the percent of stream miles with low, medium, and high levels of salinity compared to regional references. Nationally, 3% of all stream miles had high salinity conditions, whereas 82.5% had low salinity conditions.

Regionally, the highest percentage of stream miles with high salinity conditions was in the Plains and Lowlands, where 5% of all stream miles had high levels of

FIGURE 3.3

Biological condition of streams, 2006

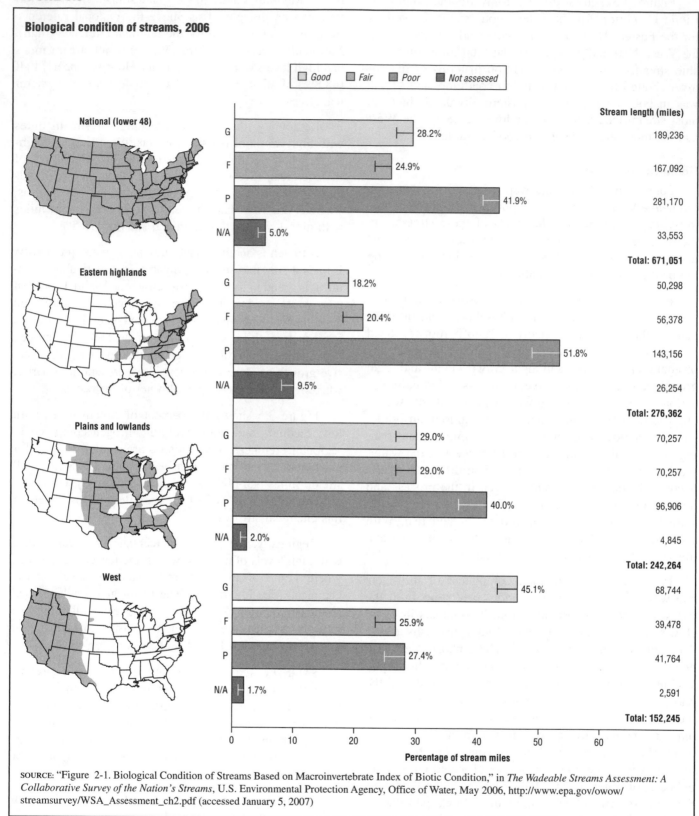

SOURCE: "Figure 2-1. Biological Condition of Streams Based on Macroinvertebrate Index of Biotic Condition," in *The Wadeable Streams Assessment: A Collaborative Survey of the Nation's Streams*, U.S. Environmental Protection Agency, Office of Water, May 2006, http://www.epa.gov/owow/streamsurvey/WSA_Assessment_ch2.pdf (accessed January 5, 2007)

salts. (See Figure 3.6.) The West had only 2.6% of its stream miles experiencing high salinity, whereas the Eastern Highlands had the least number of stream miles with high levels of this chemical stressor: 1.3%.

ACIDIFICATION. Stream acidification means that the water has become more acidic than is natural. Figure 3.7 shows that the pH of a healthy lake (or stream) is about 6.5. The pH scale shows levels of acidity. This scale is numbered from zero

FIGURE 3.4

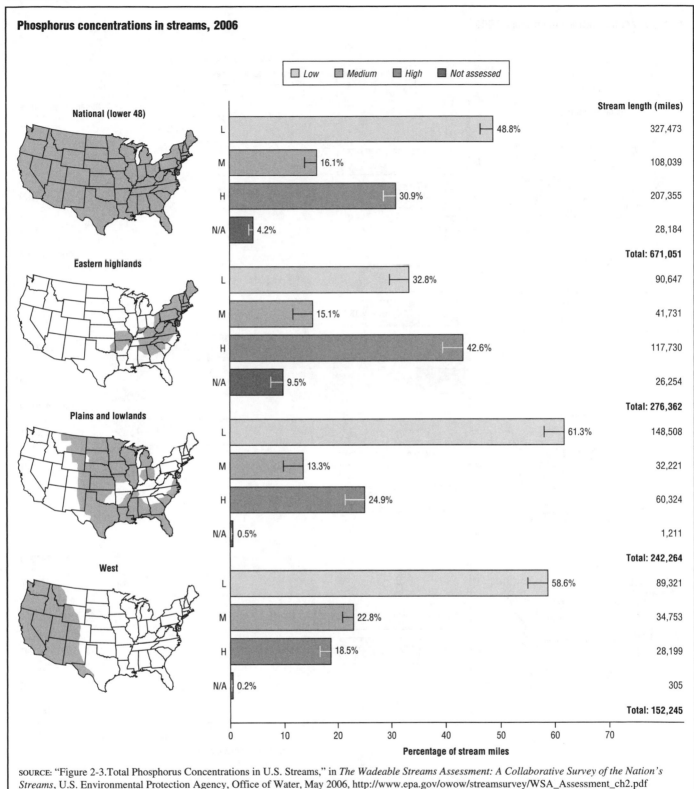

Phosphorus concentrations in streams, 2006

SOURCE: "Figure 2-3. Total Phosphorus Concentrations in U.S. Streams," in *The Wadeable Streams Assessment: A Collaborative Survey of the Nation's Streams*, U.S. Environmental Protection Agency, Office of Water, May 2006, http://www.epa.gov/owow/streamsurvey/WSA_Assessment_ch2.pdf (accessed January 5, 2007)

to fourteen, with a pH value of seven considered neutral. Values higher than seven are considered more alkaline or basic; values that are lower than seven are considered acidic. Pure, distilled water has a pH level of seven.

The pH scale is a logarithmic measure. This means that every pH drop of one is a tenfold increase in acid content. Therefore, a decrease from pH six to pH five is a tenfold increase in acidity; a drop from pH six to pH four

FIGURE 3.5

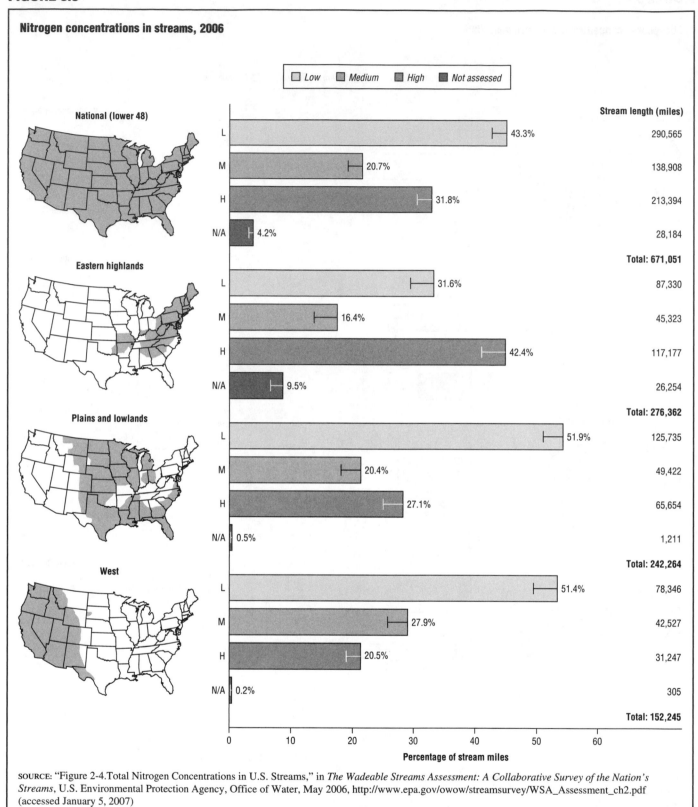

Nitrogen concentrations in streams, 2006

☐ Low   ☐ Medium   ☐ High   ■ Not assessed

Stream length (miles)

National (lower 48)
L   43.3%   290,565
M   20.7%   138,908
H   31.8%   213,394
N/A   4.2%   28,184
Total: 671,051

Eastern highlands
L   31.6%   87,330
M   16.4%   45,323
H   42.4%   117,177
N/A   9.5%   26,254
Total: 276,362

Plains and lowlands
L   51.9%   125,735
M   20.4%   49,422
H   27.1%   65,654
N/A   0.5%   1,211
Total: 242,264

West
L   51.4%   78,346
M   27.9%   42,527
H   20.5%   31,247
N/A   0.2%   305
Total: 152,245

Percentage of stream miles

SOURCE: "Figure 2-4.Total Nitrogen Concentrations in U.S. Streams," in *The Wadeable Streams Assessment: A Collaborative Survey of the Nation's Streams*, U.S. Environmental Protection Agency, Office of Water, May 2006, http://www.epa.gov/owow/streamsurvey/WSA_Assessment_ch2.pdf (accessed January 5, 2007)

is a hundredfold increase in acidity; and a drop from pH six to pH three is a thousandfold increase.

"Clean" rainfall has a pH of 5.6. It is not neutral because it is not pure water; it accumulates naturally occur-ring sulfur oxides and nitrogen oxides as it passes through the atmosphere. In comparison, acid rain (or acid deposi-tion) has a pH of about 4.2 to 4.4. The introduction of large volumes of acid deposition can over time increase the acidity of a body of water by as much as a hundredfold.

FIGURE 3.6

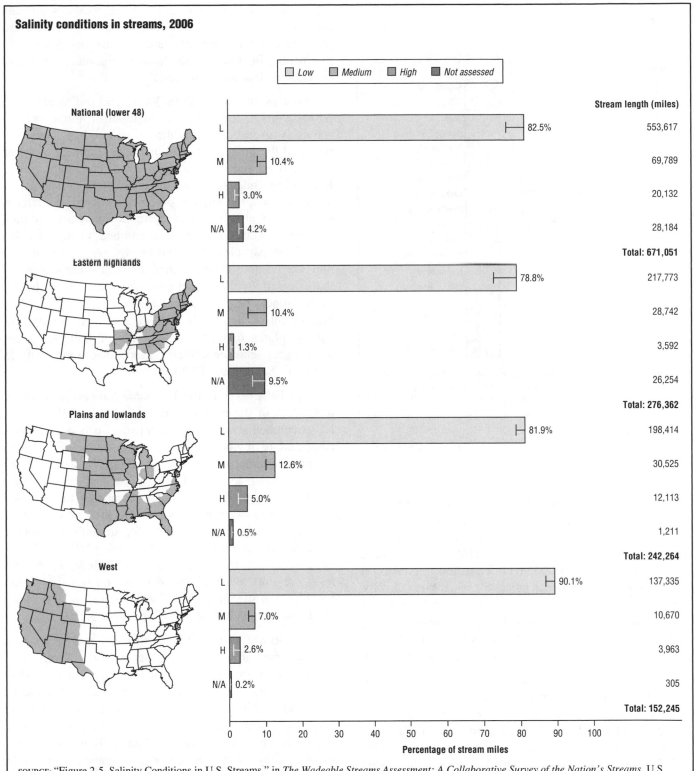

Salinity conditions in streams, 2006

| Low | Medium | High | Not assessed |

SOURCE: "Figure 2-5. Salinity Conditions in U.S. Streams," in *The Wadeable Streams Assessment: A Collaborative Survey of the Nation's Streams*, U.S. Environmental Protection Agency, Office of Water, May 2006, http://www.epa.gov/owow/streamsurvey/WSA_Assessment_ch2.pdf (accessed January 5, 2007)

One of the main components of acid deposition is sulfur dioxide from the burning of fossil fuels, mainly from auto exhaust and coal-burning power plants. (See Figure 3.8.) As sulfur dioxide reaches the atmosphere, it becomes sulfuric acid when it joins with hydrogen atoms in the air. Nitric oxide and nitric dioxide are the other major components of acid deposition. Like sulfur dioxide, these nitrogen oxides are produced from the burning

FIGURE 3.7

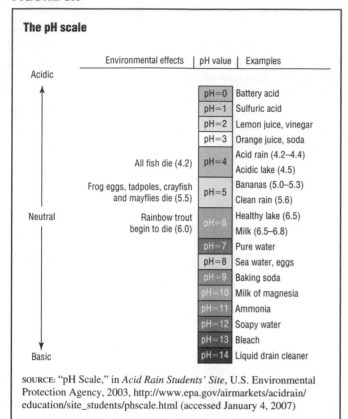

The pH scale

	Environmental effects	pH value	Examples
Acidic		pH=0	Battery acid
		pH=1	Sulfuric acid
		pH=2	Lemon juice, vinegar
		pH=3	Orange juice, soda
	All fish die (4.2)	pH=4	Acid rain (4.2–4.4)
			Acidic lake (4.5)
	Frog eggs, tadpoles, crayfish and mayflies die (5.5)	pH=5	Bananas (5.0–5.3)
			Clean rain (5.6)
Neutral	Rainbow trout begin to die (6.0)	pH=6	Healthy lake (6.5)
			Milk (6.5–6.8)
		pH=7	Pure water
		pH=8	Sea water, eggs
		pH=9	Baking soda
		pH=10	Milk of magnesia
		pH=11	Ammonia
		pH=12	Soapy water
		pH=13	Bleach
Basic		pH=14	Liquid drain cleaner

SOURCE: "pH Scale," in *Acid Rain Students' Site*, U.S. Environmental Protection Agency, 2003, http://www.epa.gov/airmarkets/acidrain/education/site_students/phscale.html (accessed January 4, 2007)

of fossil fuels. They rise into the atmosphere and oxidize in clouds to form nitric acid.

Figure 3.8 illustrates how sulfur and nitrogen oxides are carried into the air to become acid deposition. Gases and particulate matter are carried into the atmosphere, where they mix with moisture and other pollutants to form dry (aerosols, particles, and gases) and wet (fog, hail, rain, sleet, snow, dew) acid deposition. Wet deposition returns to the earth as precipitation, which enters the water body directly, percolates through the soil, or becomes runoff to nearby water bodies. Dry deposition builds up over time on all dry surfaces and is transported to water bodies in runoff during periods of precipitation or falls directly onto a water surface.

Figure 3.9 shows acidification in U.S. streams. In the *Wadeable Streams Assessment*, the EPA states that "about 2% of the nation's stream length (14,763 miles) is impacted by acidification from anthropogenic [human-related] sources. These sources include acid deposition (0.7%), acid mine drainage (0.4%), and episodic acidity due to high-runoff events (1%). Although these numbers appear relatively small, they reflect a significant impact in certain parts of the United States, particularly in the Eastern Highlands region, where 3.4% of the stream length (9,396) is impacted by acidification."

Physical Habitat Stressors

Freshwater streams are the physical habitats (natural homes) for a variety of plants and animals. The physical characteristics of a stream can be changed by human activities, and those changes can be stressors for the organisms that live there. EPA researchers assessed these physical characteristics of wadeable streams: streambed sediments, in-stream fish habitat, riparian vegetative cover, and riparian disturbance.

STREAMBED SEDIMENTS. Water and sediments drain into streams as the result of a number of human-related activities, including agriculture, road building, construction, and the grazing of farm animals. Drainage of water and sediments into a stream can affect its size and shape. In addition, the size of the sediment particles can affect the streambed. If sediment particles are large, the stream may not be able to move them downstream, so they eventually accumulate in the streambed, changing habitats. If the particles are small but excessive, the stream may not be able to move them as well. Fine sediments that are left in the streambed can begin filling in the habitat spaces between stones and boulders on the stream bottom. Suspended fine sediments can block sunlight to aquatic plants and abrade the gills of fish. All these occurrences can negatively affect macroinvertebrates and fish. (See Figure 3.10.)

The EPA notes in the *Wadeable Streams Assessment* that "25% of the nation's stream length (167,092 miles) has streambed sediment characteristics in poor condition compared to regional reference condition.... Streambed sediment characteristics are rated fair in 20% of the nation's stream length (132,197 miles) and good in 50% of stream length (336,197 miles) compared to reference condition. The two regions with the greatest percentage of stream length in poor condition for streambed sediment characteristics are the Eastern Highlands (28%, or 77,381 miles) and the Plains and Lowlands (26%, or 63,958 miles) regions, whereas the West region has the lowest percentage of stream length (17%, or 26,522 miles) in poor condition for this indicator." (See Figure 3.11.)

IN-STREAM FISH HABITAT. Streams and rivers that have diverse and complex habitats support a diversity of fish and macroinvertebrates. Such habitats include undercut banks with exposed tree roots, brush and large pieces of wood within the stream, and boulders within the stream and at the stream bank. Cover from overhanging vegetation also affects stream habitats. When humans use streams, they often change these complex habitats to simpler ones, which often results in a reduction in the diversity of the organisms living there.

Figure 3.12 shows that 19.5% of the stream miles across the nation had poor in-stream habitat conditions. About one-fourth (24.9%) had fair conditions, and about half (51.5%) had good conditions. The highest percentage of stream miles with poor in-stream habitat conditions was in the Plains and Lowlands, with 37% of stream miles rated poor. The West was next, with 12.3% of

FIGURE 3.8

Transported air pollutants: emissions to effects

Transported pollutants result from emissions of three pollutants: sulfur dioxide, nitrogen oxides, and hydrocarbons. As these pollutants are carried away from their sources, they form a complex "pollutant mix" leading to acid deposition, ozone, and airborne fine particles. These transported air pollutants pose risks to surface waters, forests, crops, materials, visibility, and human health.

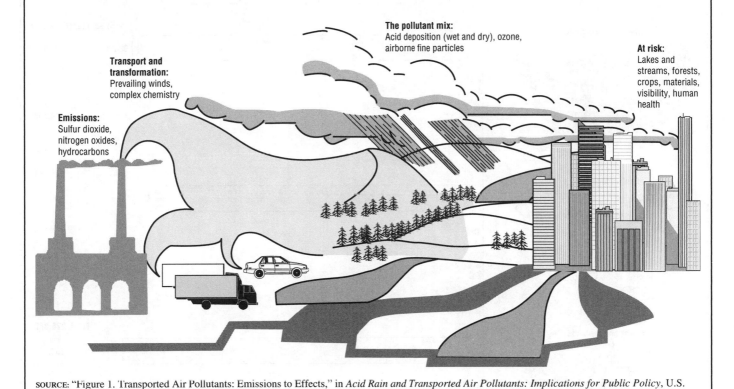

The pollutant mix:
Acid deposition (wet and dry), ozone, airborne fine particles

Transport and transformation:
Prevailing winds, complex chemistry

At risk:
Lakes and streams, forests, crops, materials, visibility, human health

Emissions:
Sulfur dioxide, nitrogen oxides, hydrocarbons

SOURCE: "Figure 1. Transported Air Pollutants: Emissions to Effects," in *Acid Rain and Transported Air Pollutants: Implications for Public Policy*, U.S. Congress, Office of Technology Assessment, June 1984, http://www.wws.princeton.edu/ota/disk3/1984/8401/8401.PDF (accessed January 23, 2007)

stream miles rated poor, and the Eastern Highlands had only 8.2% of stream miles rated poor. Nonetheless, the West had the greatest number of stream miles with in-stream habitats rated good: 66.4%.

RIPARIAN VEGETATIVE COVER. The word *riparian* means the banks of a body of water, such as a river or stream. In the *Wadeable Streams Assessment*, riparian vegetative cover refers to the amount and type of vegetation growing on or next to stream banks; it is an indicator of the health of a stream. Complex, multilayered riparian vegetation helps maintain the health of the stream by reducing runoff from the surrounding land, preventing stream bank erosion, supplying shade, and providing food and habitats in the form of leaf litter and large wood. As with in-stream habitats, riparian coverage is often changed, or simplified, by humans. EPA researchers assessed the ground layer, woody shrubs, and canopy trees of the riparian cover of streams.

Figure 3.13 shows that 19.3% of the wadeable stream miles nationally were in poor condition "due to severely simplified riparian vegetation," 28.3% were in fair condition, and 47.6% were in good condition. Regionally, the Plains and Lowlands had the greatest percentage of stream miles in poor condition (26%) with respect to

riparian coverage. The Eastern Highlands followed with 17.6% in poor condition, and the West had the least number of stream miles in poor condition: 12.2%.

RIPARIAN DISTURBANCE. As mentioned earlier, human activities can change or disturb the riparian vegetative cover. The closer potentially disturbing human activities take place to the stream bank, the more likely they are to cause riparian disturbance. To determine riparian human disturbance, EPA researchers tallied eleven forms of human activities and disturbances along sections of streams and weighted them according to how close they were to the streams.

Figure 3.14 shows that nationally one-fourth (25.5%) of stream length had high riparian disturbance when compared to reference sites. Nearly half (46.8%) had fair riparian disturbance, and one-fourth (23.6%) had low riparian disturbance. The EPA notes in the *Wadeable Streams Assessment* that "one of the striking findings [was] the widespread distribution of intermediate levels of riparian disturbance," both nationally and regionally. Furthermore, the EPA states, "It is worth noting that for the nation and the three regions, the amount of stream length with good riparian vegetative cover was significantly greater than the amount of stream length with low levels of human

FIGURE 3.9

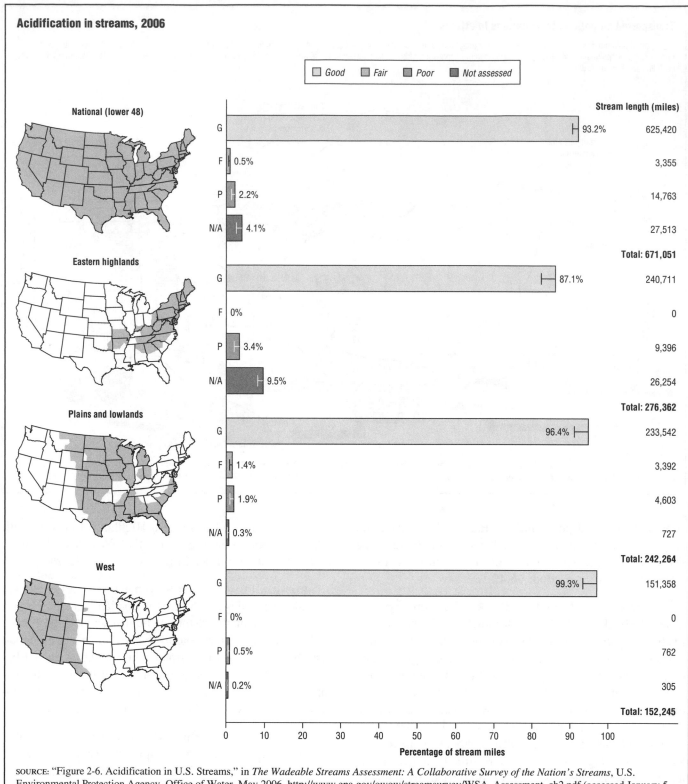

Acidification in streams, 2006

Good | Fair | Poor | Not assessed

National (lower 48)

Stream length (miles)

G 93.2% — 625,420

F 0.5% — 3,355

P 2.2% — 14,763

N/A 4.1% — 27,513

Total: 671,051

Eastern highlands

G 87.1% — 240,711

F 0% — 0

P 3.4% — 9,396

N/A 9.5% — 26,254

Total: 276,362

Plains and lowlands

G 96.4% — 233,542

F 1.4% — 3,392

P 1.9% — 4,603

N/A 0.3% — 727

Total: 242,264

West

G 99.3% — 151,358

F 0% — 0

P 0.5% — 762

N/A 0.2% — 305

Total: 152,245

Percentage of stream miles

SOURCE: "Figure 2-6. Acidification in U.S. Streams," in *The Wadeable Streams Assessment: A Collaborative Survey of the Nation's Streams*, U.S. Environmental Protection Agency, Office of Water, May 2006, http://www.epa.gov/owow/streamsurvey/WSA_Assessment_ch2.pdf (accessed January 5, 2007)

disturbance in the riparian zone. This finding warrants additional investigation, but suggests that land managers and property owners are protecting and maintaining healthy riparian vegetation buffers, even along streams where disturbance from roads, agriculture, and grazing is widespread."

Relative Extent and Relative Risk of Stressors

Figure 3.15 shows a list of stressors caused by human activity, the relative extent to which each affected the freshwater streams of the United States in 2006, and the relative risk they posed to macroinvertebrates.

FIGURE 3.10

Effects of siltation in rivers and streams

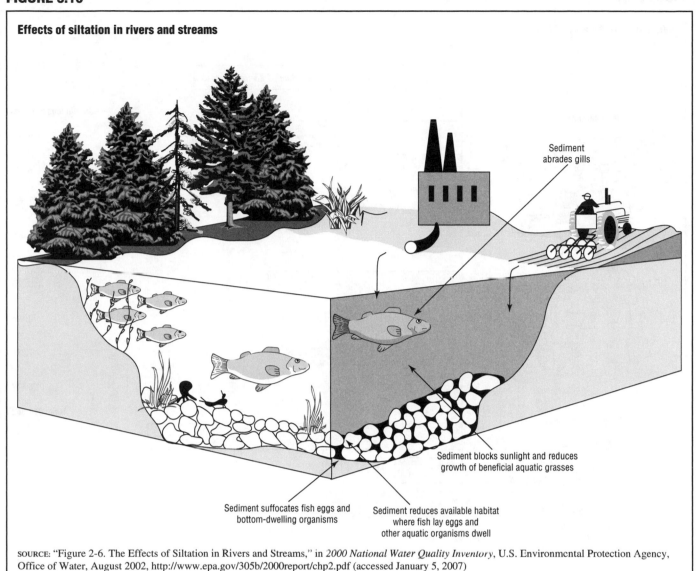

SOURCE: "Figure 2-6. The Effects of Siltation in Rivers and Streams," in *2000 National Water Quality Inventory*, U.S. Environmental Protection Agency, Office of Water, August 2002, http://www.epa.gov/305b/2000report/chp2.pdf (accessed January 5, 2007).

The bar graph on the left shows what percentage of stream length nationally each stressor affects and what its relationship is to the other stressors. Each stressor is ranked according to the proportion of stream length that was in poor condition nationally. Excessive nitrogen levels affected the greatest percentage of stream miles at 31.8%. Phosphorus, the other plant nutrient assessed, was a close second, affecting 30.9% of stream miles nationally. Excess salinity and acidification affected the least percentage of stream miles, 2.9% and 2.2%, respectively.

The bar graph on the right shows the national relative risk values for each stressor. A relative risk value of 1 means no association between the stressor and the biologic health of wadeable stream macroinvertebrates. Thus, acidification, with a relative risk value of 1, does not pose a risk to stream macroinvertebrates. Values greater than 1 suggest a positive association; the higher the number, the greater the risk. As such, the poor con-

dition of streambed sediments poses the greatest relative risk to stream macroinvertebrates of all the stressors assessed. Excessive phosphorus and nitrogen pose the next greatest relative risks, respectively.

WATER QUALITY OF THE NATION'S LAKES

The older EPA water quality reports were set up quite differently from the *Wadeable Streams Assessment*, which is typical of the new reports published by the EPA on water quality. In the *2000 National Water Quality Inventory* report, a use was designated for surface water bodies in each state. The state then established water quality numeric and narrative criteria to protect each use. More than one designated use was frequently assigned to a water body. Most water bodies were designated for recreation, drinking water use, and protection of aquatic life.

The *2000 National Water Quality Inventory* assessed 43% of the nation's 40.6 million acres of lakes, ponds,

FIGURE 3.11

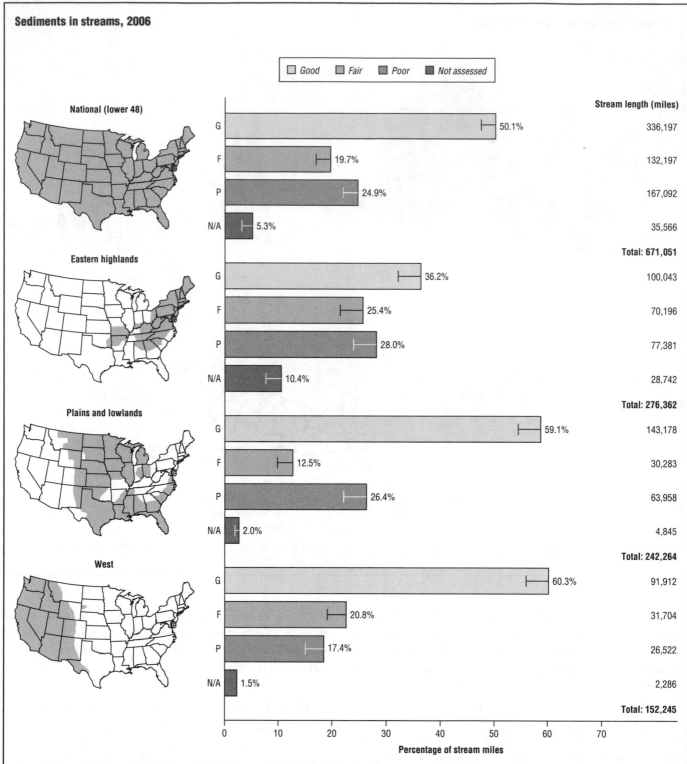

Sediments in streams, 2006

□ Good ▨ Fair ▨ Poor ■ Not assessed

National (lower 48)

	Percentage	Stream length (miles)
G	50.1%	336,197
F	19.7%	132,197
P	24.9%	167,092
N/A	5.3%	35,566
		Total: 671,051

Eastern highlands

	Percentage	
G	36.2%	100,043
F	25.4%	70,196
P	28.0%	77,381
N/A	10.4%	28,742
		Total: 276,362

Plains and lowlands

	Percentage	
G	59.1%	143,178
F	12.5%	30,283
P	26.4%	63,958
N/A	2.0%	4,845
		Total: 242,264

West

	Percentage	
G	60.3%	91,912
F	20.8%	31,704
P	17.4%	26,522
N/A	1.5%	2,286
		Total: 152,245

Percentage of stream miles

SOURCE: "Figure 2-7. Streambed Sediments in U.S. Streams," in *The Wadeable Streams Assessment: A Collaborative Survey of the Nation's Streams*, U.S. Environmental Protection Agency, Office of Water, May 2006, http://www.epa.gov/owow/streamsurvey/WSA_Assessment_ch2.pdf (accessed January 5, 2007)

and reservoirs. Forty-seven percent were found to fully support their designated uses. (See Figure 3.16.) However, 8% of the lake acres were threatened. Of the lakes assessed, 45% could only partially support their designated uses.

Leading Pollutants/Stressors in Lakes, Ponds, and Reservoirs

A lake's water quality reflects the condition and management of its watershed. In 2000 elevated levels of

FIGURE 3.12

Condition of in-stream fish habitats, 2006

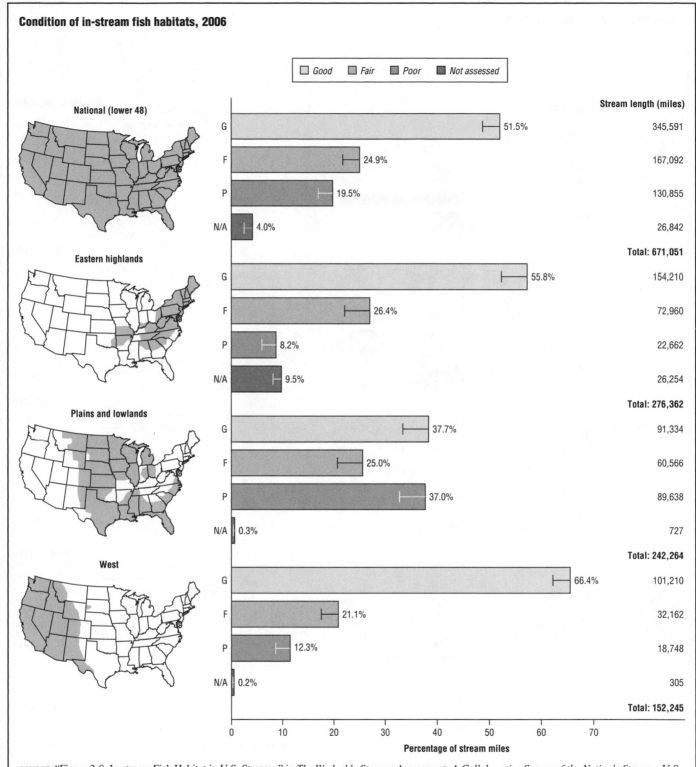

☐ Good ☐ Fair ☐ Poor ■ Not assessed

National (lower 48)

		Stream length (miles)
G	51.5%	345,591
F	24.9%	167,092
P	19.5%	130,855
N/A	4.0%	26,842
		Total: 671,051

Eastern highlands

G	55.8%	154,210
F	26.4%	72,960
P	8.2%	22,662
N/A	9.5%	26,254
		Total: 276,362

Plains and lowlands

G	37.7%	91,334
F	25.0%	60,566
P	37.0%	89,638
N/A	0.3%	727
		Total: 242,264

West

G	66.4%	101,210
F	21.1%	32,162
P	12.3%	18,748
N/A	0.2%	305
		Total: 152,245

Percentage of stream miles

SOURCE: "Figure 2-8. In-stream Fish Habitat in U.S. Streams," in *The Wadeable Streams Assessment: A Collaborative Survey of the Nation's Streams*, U.S. Environmental Protection Agency, Office of Water, May 2006, http://www.epa.gov/owow/streamsurvey/WSA_Assessment_ch2.pdf (accessed January 5, 2007)

plant nutrients (phosphorus and nitrogen) were identified as the most common stressors, contributing to 50% of the impaired water quality in lakes. Figure 3.17 shows the top stressors and the percentage of impaired lake acres affected by each.

Metals were the second most prevalent stressor, affecting 42% of the impaired lake acres. This finding was caused mostly by the widespread detection of mercury in fish tissue. Because it is difficult to measure mercury in water, and because mercury readily accumulates in tissue

FIGURE 3.13

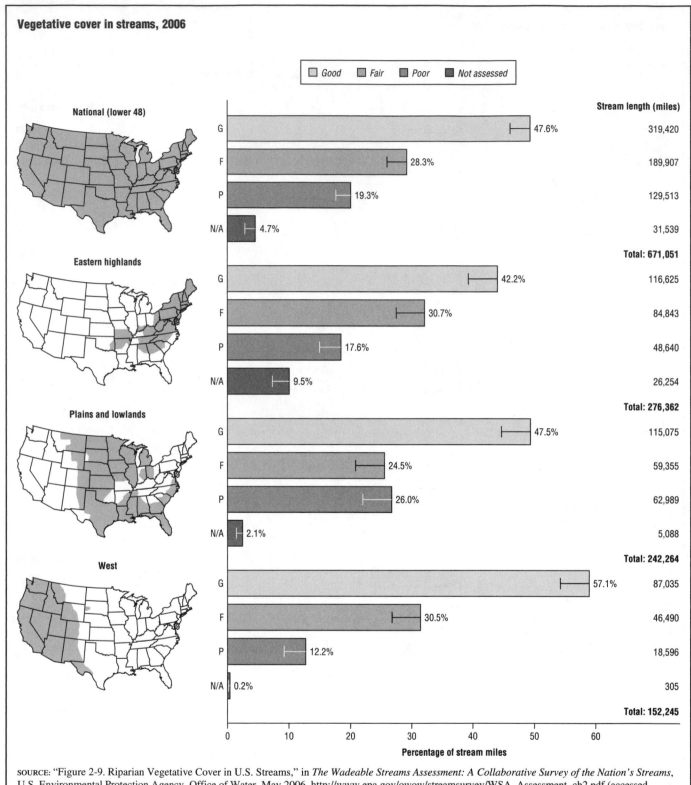

Vegetative cover in streams, 2006

Legend: ☐ Good ▨ Fair ▨ Poor ▉ Not assessed

National (lower 48)

		Stream length (miles)
G	47.6%	319,420
F	28.3%	189,907
P	19.3%	129,513
N/A	4.7%	31,539
		Total: 671,051

Eastern highlands

G	42.2%	116,625
F	30.7%	84,843
P	17.6%	48,640
N/A	9.5%	26,254
		Total: 276,362

Plains and lowlands

G	47.5%	115,075
F	24.5%	59,355
P	26.0%	62,989
N/A	2.1%	5,088
		Total: 242,264

West

G	57.1%	87,035
F	30.5%	46,490
P	12.2%	18,596
N/A	0.2%	305
		Total: 152,245

Percentage of stream miles (0, 10, 20, 30, 40, 50, 60)

SOURCE: "Figure 2-9. Riparian Vegetative Cover in U.S. Streams," in *The Wadeable Streams Assessment: A Collaborative Survey of the Nation's Streams*, U.S. Environmental Protection Agency, Office of Water, May 2006, http://www.epa.gov/owow/streamsurvey/WSA_Assessment_ch2.pdf (accessed January 5, 2007)

(bioaccumulates), most states measure mercury contamination using fish tissue samples. Mercury generally enters the water from the air, often released from the smokestacks of power-generating facilities, waste incinerators, and other sources.

The third most common pollutant of lakes reported in the EPA's 2000 inventory was siltation or sedimentation. Nine percent of the lakes assessed in the report were shown to have been impaired by siltation, making this pollutant responsible for 21% of the lake acres designated as impaired. (Figure 3.17.)

FIGURE 3.14

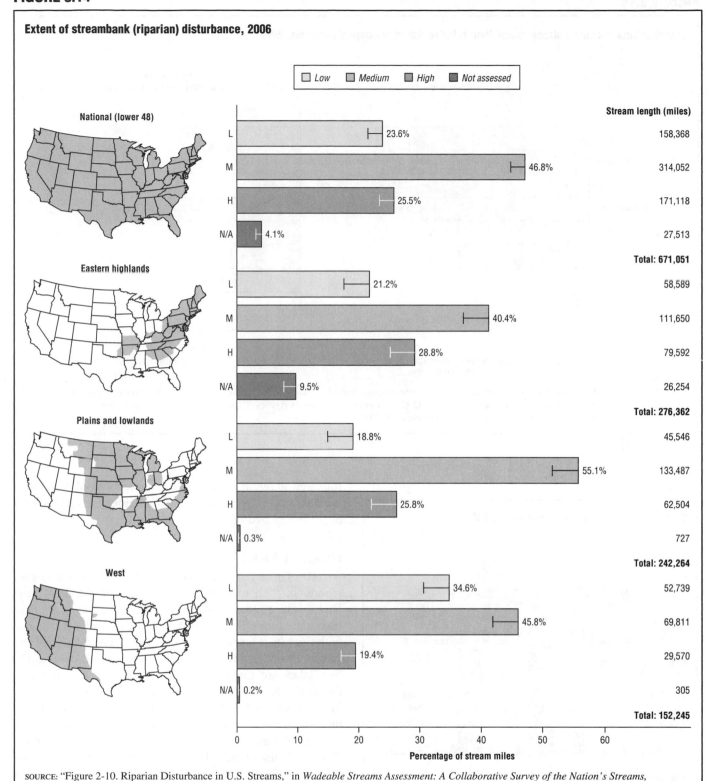

Extent of streambank (riparian) disturbance, 2006

Legend: ☐ Low ☐ Medium ▨ High ■ Not assessed

National (lower 48)

		Stream length (miles)
L	23.6%	158,368
M	46.8%	314,052
H	25.5%	171,118
N/A	4.1%	27,513
		Total: 671,051

Eastern highlands

L	21.2%	58,589
M	40.4%	111,650
H	28.8%	79,592
N/A	9.5%	26,254
		Total: 276,362

Plains and lowlands

L	18.8%	45,546
M	55.1%	133,487
H	25.8%	62,504
N/A	0.3%	727
		Total: 242,264

West

L	34.6%	52,739
M	45.8%	69,811
H	19.4%	29,570
N/A	0.2%	305
		Total: 152,245

Percentage of stream miles

SOURCE: "Figure 2-10. Riparian Disturbance in U.S. Streams," in *Wadeable Streams Assessment: A Collaborative Survey of the Nation's Streams,* U.S. Environmental Protection Agency, Office of Water, May 2006, http://www.epa.gov/owow/streamsurvey/WSA_Assessment_ch2.pdf (accessed January 5, 2007)

Sources of Pollutants/Stressors in Lakes

As in the case of streams, agricultural runoff was a significant source of pollution for lakes, affecting 41% of impaired lake acres in 2000. (See Figure 3.18.) Pasture grazing and both irrigated and nonirrigated crop produc-

tion were the leading sources of agricultural impairments to lake water quality.

The second most commonly found cause of lake impairment was hydrologic modifications. (See Figure 3.18.) These modifications, resulting from regulation of the flow of water,

FIGURE 3.15

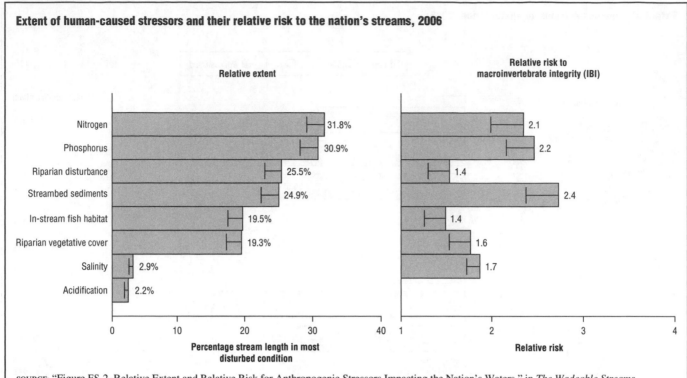

Extent of human-caused stressors and their relative risk to the nation's streams, 2006

Relative extent

Nitrogen	31.8%
Phosphorus	30.9%
Riparian disturbance	25.5%
Streambed sediments	24.9%
In-stream fish habitat	19.5%
Riparian vegetative cover	19.3%
Salinity	2.9%
Acidification	2.2%

0   10   20   30   40
Percentage stream length in most
disturbed condition

Relative risk to
macroinvertebrate integrity (IBI)

	2.1
	2.2
	1.4
	2.4
	1.4
	1.6
	1.7

1   2   3   4
Relative risk

SOURCE: "Figure ES-2. Relative Extent and Relative Risk for Anthropogenic Stressors Impacting the Nation's Waters," in *The Wadeable Streams Assessment: A Collaborative Survey of the Nation's Streams*, U.S. Environmental Protection Agency, Office of Water, May 2006, http://www.epa.gov/owow/streamsurvey/WSA_Assessment_intro.pdf (accessed January 5, 2007)

FIGURE 3.16

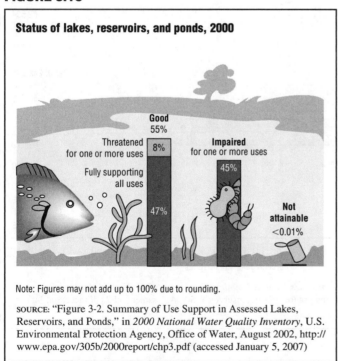

Status of lakes, reservoirs, and ponds, 2000

Good
55%

Threatened
for one or more uses
8%

Fully supporting
all uses
47%

Impaired
for one or more uses
45%

Not
attainable
<0.01%

Note: Figures may not add up to 100% due to rounding.

SOURCE: "Figure 3-2. Summary of Use Support in Assessed Lakes, Reservoirs, and Ponds," in *2000 National Water Quality Inventory*, U.S. Environmental Protection Agency, Office of Water, August 2002, http://www.epa.gov/305b/2000report/chp3.pdf (accessed January 5, 2007)

impaired acres. A nearly equal percentage of lake acres were degraded by urban runoff and storm sewers as were degraded by hydrologic modifications.

GREAT LAKES

The Great Lakes Environmental Research Laboratory reports in "About Our Great Lakes: Great Lakes Basin Facts" (June 18, 2004, http://www.glerl.noaa.gov/pr/our lakes/facts.html) that the Great Lakes basin, which is shared with Canada, is home to thirty-five million people. The lakes provide drinking water for about forty million people. The five lakes are the largest surface area of freshwater in the world, at ninety-five thousand square miles. The water in the Great Lakes accounts for 90% of all the freshwater in the United States. The total shoreline of the Great Lakes in the United States and Canada, a "fourth seacoast," is more than ten thousand miles long and equal to about one-quarter of the earth's circumference. International shipping on the Lakes annually transports two hundred million tons of cargo, and sport fishing contributes $4 billion to the economy.

The prosperity of the Great Lakes region, however, has taxed its ecological health. Urban and industrial discharges, agricultural and forestry activity, development of recreation facilities, poor waste disposal practices, invasive species, and habitat degradation have all contributed to ecosystem decline. Despite these problems, the watershed still contains many ecologically rich areas. In "About Our Great Lakes:

dredging, and construction of dams, degraded 8% of the assessed lake, pond, and reservoir acres and 18% of the

FIGURE 3.17

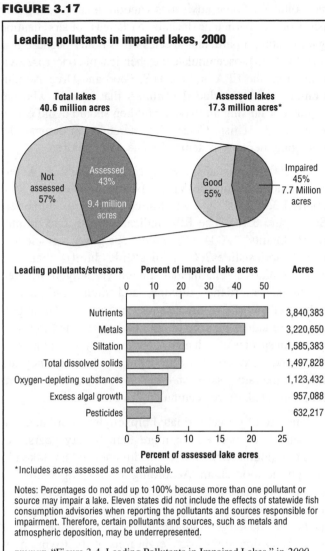

Leading pollutants in impaired lakes, 2000

Total lakes
40.6 million acres

Assessed lakes
17.3 million acres*

Not assessed 57%

Assessed 43%
9.4 million acres

Good 55%

Impaired 45%
7.7 Million acres

Leading pollutants/stressors	Percent of impaired lake acres	Acres
Nutrients		3,840,383
Metals		3,220,650
Siltation		1,585,383
Total dissolved solids		1,497,828
Oxygen-depleting substances		1,123,432
Excess algal growth		957,088
Pesticides		632,217

Percent of assessed lake acres

*Includes acres assessed as not attainable.

Notes: Percentages do not add up to 100% because more than one pollutant or source may impair a lake. Eleven states did not include the effects of statewide fish consumption advisories when reporting the pollutants and sources responsible for impairment. Therefore, certain pollutants and sources, such as metals and atmospheric deposition, may be underrepresented.

SOURCE: "Figure 3-4. Leading Pollutants in Impaired Lakes," in *2000 National Water Quality Inventory*, U.S. Environmental Protection Agency, Office of Water, August 2002, http://www.epa.gov/305b/2000 report/chp3.pdf (accessed January 5, 2007)

FIGURE 3.18

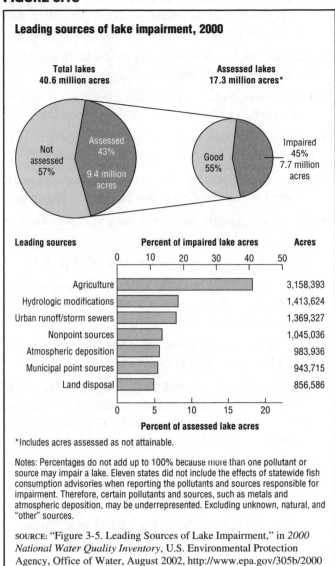

Leading sources of lake impairment, 2000

Total lakes
40.6 million acres

Assessed lakes
17.3 million acres*

Not assessed 57%

Assessed 43%
9.4 million acres

Good 55%

Impaired 45%
7.7 million acres

Leading sources	Percent of impaired lake acres	Acres
Agriculture		3,158,393
Hydrologic modifications		1,413,624
Urban runoff/storm sewers		1,369,327
Nonpoint sources		1,045,036
Atmospheric deposition		983,936
Municipal point sources		943,715
Land disposal		856,586

Percent of assessed lake acres

*Includes acres assessed as not attainable.

Notes: Percentages do not add up to 100% because more than one pollutant or source may impair a lake. Eleven states did not include the effects of statewide fish consumption advisories when reporting the pollutants and sources responsible for impairment. Therefore, certain pollutants and sources, such as metals and atmospheric deposition, may be underrepresented. Excluding unknown, natural, and "other" sources.

SOURCE: "Figure 3-5. Leading Sources of Lake Impairment," in *2000 National Water Quality Inventory*, U.S. Environmental Protection Agency, Office of Water, August 2002, http://www.epa.gov/305b/2000 report/chp3.pdf (accessed January 5, 2007)

Ecology" (June 2, 2004, http://www.glerl.noaa.gov/pr/ourlakes/ecology.html), the Great Lakes Environmental Research Laboratory indicates that approximately thirty-five hundred species of plants and animals inhabit the Great Lakes basin.

Great Lakes Water Quality Agreement

In 1972 the United States and Canada entered into the Great Lakes Water Quality Agreement, which is a worldwide model for cooperative environmental protection and natural resource management. The agreement imposes reporting requirements on both of its member countries, and in an attempt to meet these requirements a conference series was established. The conferences, which are held every two years, are called the State of the Lakes Ecosystem Conference (SOLEC). The first such conference was convened in 1994 and the most recent was in late 2006.

The SOLEC meetings are designed as a venue for scientists and policy makers to share information about the state of the Great Lakes ecosystem. The focus is on assessing and sharing information about the results of Great Lakes programs and studies. In the year following each conference, the United States and Canada prepare a report that presents the findings accumulated at the SOLEC.

The report published after SOLEC 2004, *State of the Great Lakes 2005* (2005, http://binational.net/solec/English/SOLEC%202004/Tagged%20PDFs/SOGL%202005%20Report/English%20Version/Complete%20Report.pdf), presents the following mixed news about the chemical, physical, and biological integrity of the waters of the Great Lakes basin ecosystem. Some of the good features identified include:

- Persistent toxic substances are continuing to decline.

- The Great Lakes are a good source for treated drinking water.

- Total forested land in the Great Lakes basin appears to have increased in recent decades. Approximately 50% of the Great Lakes basin is covered by forest.

- Bald eagles are continuing to nest and fledge along the Great Lakes shorelines.

- Lake trout stocks in Lake Superior have remained self-sustaining.

- Natural reproduction of lake trout is evident in Lake Ontario and in isolated areas of Lake Huron.

- Mayfly (*Hexagenia*) populations have partially recovered in western Lake Erie and in the Bay of Quinte, Lake Ontario.

- Phosphorus targets have been met in Lakes Ontario, Huron, Michigan and Superior.

Some of the negative features identified include:

- Nonnative species are a significant threat to the ecosystem and continue to enter the Great Lakes (aquatic and terrestrial species).

- Scud (*Diporeia*) populations continue to decline in Lakes Michigan, Ontario, and Huron.

- Type E botulism outbreaks, resulting in the deaths of fish and fish-eating birds, have recently been detected in a few locations along the Lake Ontario shoreline, and minor outbreaks are continuing in Lake Erie.

- Groundwater resources are being negatively impacted by development, withdrawal and agricultural drainage.

- Long-range atmospheric transport is a continuing source of contaminants to the Great Lakes basin.

- Native mussel populations continue to be decimated as a result of invasive zebra mussels.

- Land use changes in [favor] of urbanization along the shoreline continue to threaten natural habitats in the Great Lakes and St. Lawrence River ecosystems.

- Some species of amphibians and wetland-dependent birds are showing declines in population numbers—in part due to wetland habitat conditions.

- Phosphorus levels are still above guidelines in Lake Erie.

FISH ADVISORIES

When fish or shellfish in particular locations contain harmful levels of pollutants, the state issues advisories to recreational fishermen against eating the fish. Commercial fishing is usually banned. Since 1993 the EPA has compiled these advisories annually and made them available to the public. Fish advisories are advice to limit or avoid eating certain fish.

Figure 3.19 shows the number of advisories against eating fish or wildlife reported by the states to the EPA in 2004. These advisories are specific as to location, species,

and pollutant. Some advisories caution against eating any fish from a particular location, whereas others caution against eating a particular species of fish only because it is more likely to bioaccumulate the chemical of concern. Advisories from the EPA and the U.S. Food and Drug Administration in 2004 included warnings that women who are pregnant or nursing and young children should avoid eating certain kinds of fish. Consuming mercury can damage the developing nervous systems of babies and children.

In 2004 fifty states, the District of Columbia, American Samoa, Guam, the Virgin Islands, and Puerto Rico reported 3,221 fish and wildlife consumption advisories. (See Figure 3.19.) The EPA indicates in "Bioaccumulative Pollutants" (2004, http://www.epa.gov/waterscience/fish/advisories/slides2004-1_files/slide3.html) that the bioaccumulative chemicals—mercury, polychlorinated biphenyls, chlordane, dioxins, and dichloro-diphenyl-trichloroethane (DDT)—cause most advisories. These pollutants are called bioaccumulative because when ingested by certain species of fish and waterfowl they are not metabolized and excreted from the organism. Instead, they are stored in the fatty tissues and remain there. As more chemical is ingested, more accumulates in the organism.

The use of polychlorinated biphenyls, chlordane, and DDT has been banned for more than twenty years, yet these compounds persist in the sediments and are taken in through the food chain. According to "Fishing Warnings up Due to Mercury Pollution—EPA" (Reuters News Service, August 25, 2004), U.S. coal-burning power plants are the largest source of mercury in the United States, releasing about forty-eight tons of the toxin annually. Federal standards require a 70% reduction in mercury emissions by 2018, but some states are passing legislation to reduce emissions much sooner.

Recreational Water-Associated Outbreaks

In the *2000 National Water Quality Inventory*, four states reported that they had no record of recreation restrictions reported to them by their respective health departments, and thirteen states and tribes identified over two hundred sites where recreation was restricted at least once during the reporting cycle. Local health departments closed many of those sites more than once. Pathogens (disease-causing organisms) caused most of the restrictions. State reporting on recreational restrictions, such as beach closures, is often incomplete because agencies rely on local health departments to voluntarily monitor and report beach closures.

Eric J. Dziuban et al., in "Surveillance for Waterborne Disease and Outbreaks Associated with Recreational Water—United States, 2003–2004" (*Morbidity and Mortality Weekly Report*, December 22, 2006), list the incidence of disease outbreaks caused by recreational water contact. During this two-year period twenty-six states and

FIGURE 3.19

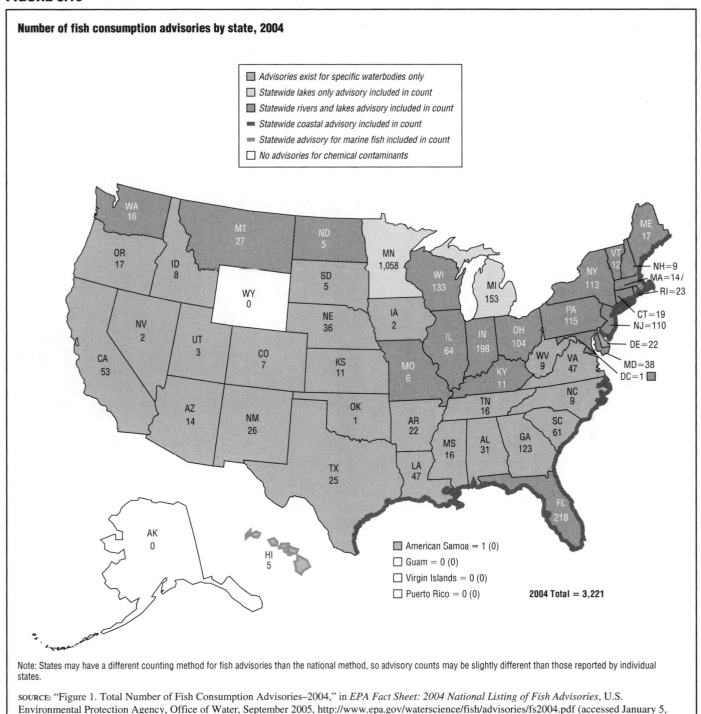

Number of fish consumption advisories by state, 2004

Advisories exist for specific waterbodies only
Statewide lakes only advisory included in count
Statewide rivers and lakes advisory included in count
Statewide coastal advisory included in count
Statewide advisory for marine fish included in count
No advisories for chemical contaminants

WA 16
MT 27
ND 5
ME 17
OR 17
ID 8
MN 1,058
VT 12
NH=9
WY 0
SD 5
WI 133
NY 113
MA=14/
NV 2
NE 36
IA 2
MI 153
RI=23
UT 3
CO 7
IL 64
IN 198
OH 104
PA 115
CT=19
NJ=110
CA 53
KS 11
WV 9
VA 47
DE=22
AZ 14
NM 26
OK 1
MO 6
KY 11
MD=38
DC=1
TN 16
NC 9
AR 22
SC 61
MS 16
AL 31
GA 123
TX 25
LA 47
FL 218
AK 0
HI 5

American Samoa = 1 (0)
Guam = 0 (0)
Virgin Islands = 0 (0)
Puerto Rico = 0 (0)   **2004 Total = 3,221**

Note: States may have a different counting method for fish advisories than the national method, so advisory counts may be slightly different than those reported by individual states.

SOURCE: "Figure 1. Total Number of Fish Consumption Advisories–2004," in *EPA Fact Sheet: 2004 National Listing of Fish Advisories*, U.S. Environmental Protection Agency, Office of Water, September 2005, http://www.epa.gov/waterscience/fish/advisories/fs2004.pdf (accessed January 5, 2007)

the territory of Guam reported sixty-two outbreaks involving nearly twenty-seven hundred people. Of the sixty-two recreational waterborne disease outbreaks reported, thirty involved gastroenteritis. There were fifteen such outbreaks in 1999 and twenty-one in 2000. Figure 3.20 shows the number of waterborne disease outbreaks due to recreational water use annually from 1978 to 2004, with a breakdown by illness.

As part of the Beaches Environmental Assessment and Coastal Health (BEACH) Act of 2000, Congress directed the EPA to develop a new set of guidelines for recreational water based on new water quality indicators. Beginning in 2003 the EPA was required to conduct a series of epidemiologic studies at recreational freshwater and marine beaches. These studies were to be used to help in the development of the new guidelines for recreational water.

FIGURE 3.20

Number of waterborne-disease outbreaks associated with recreational water use, by illness,1978–2004

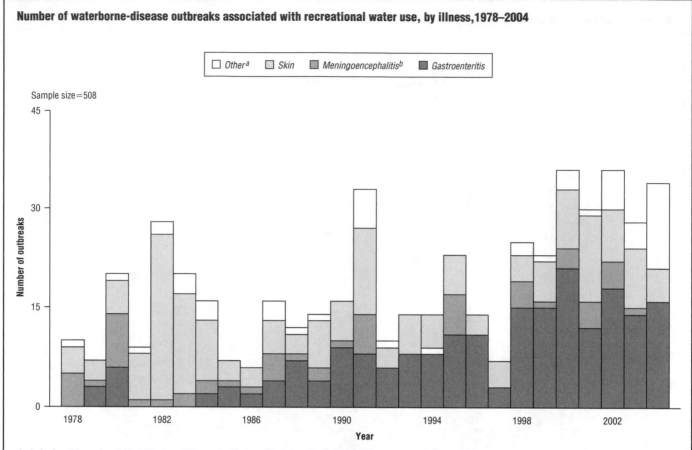

aIncludes keratitis, conjunctivitis, otitis, bronchitis, meningitis, hepatitis, leptospirosis, Pontiac fever, acute respiratory illness, and combined illnesses.
bAlso includes data from report of ameba infections.

SOURCE: Eric J. Dziuban, et al., "Figure 9. Number of Recreational Water-Associated Outbreaks (n=508), by Year and Illness—United States, 1978–2004," in "Surveillance for Waterborne Disease and Outbreaks Associated with Recreational Water—United States, 2003–2004," *Morbidity and Mortality Weekly Report, Surveillance Summaries*, vol. 55, no. SS-12, December 22, 2006, http://www.cdc.gov/mmwr/pdf/ss/ss5512.pdf (accessed January 5, 2007). Meningoencephalitis data includes data from G.S. Visvesvara and J.K. Stehr-Green, "Epidemiology of Free-Living Ameba Infections," *Journal of Protozoology*, vol. 37, 1990.

The first report, *Implementing the BEACH Act of 2000: Report to Congress* (http://www.epa.gov/waterscience/beaches/report/full-rtc.pdf), was published in October 2006 by the EPA. The EPA summarizes the following achievements:

1. States have significantly improved their assessment and monitoring of beaches; the number of monitored beaches has increased from about 1,000 in 1997 to more than 3,500 out of approximately 6,000 beaches, as identified to EPA by the states for the 2004 swimming season.

2. EPA has strengthened water quality standards throughout all the coastal recreation waters in the United States; the number of coastal and Great Lakes states with up-to-date water quality criteria has increased from 11 in 2000 to 35 in 2004.

3. EPA has improved public access to data on beach advisories and closings by improving its electronic system for beach data collection and delivery systems; the system is known as "eBeaches." The public can view the beach information at http://oaspub .epa.gov/beacon/beacon_national_page.main.

4. EPA is working to improve pollution control efforts that reduce potential adverse health effects at beaches. EPA's Strategic Plan and recent National Water Program Guidance describe these actions to coordinate assessment of problems affecting beaches and to reduce pollution . . . .

5. EPA is conducting research to develop new or revised water quality criteria and more rapid methods for assessing water quality at beaches so that results can be made available in hours rather than days. Quicker tests will allow beach managers to make faster decisions about the safety of beach waters and thus help reduce the risk of illness among beachgoers.

CHAPTER 4
GROUNDWATER

A VAST HIDDEN RESOURCE

Water lies beneath almost every part of the earth's surface—mountains, plains, and deserts—but underground water is not always easy to find, and, once found, it may not be readily accessible. Groundwater may lie close to the surface, as in a marsh, or it may occur many hundreds of feet below the surface, as in some dry areas of the nation's West.

People have known about the presence of groundwater since ancient times, but it is only recently that geologists have learned how to gauge the quantity of groundwater and have begun to estimate its vast potential for use. The U.S. Geological Survey (USGS) states in "Groundwater" (2006, http://capp.water.usgs.gov/GIP/gw_gip/gw_a.htrnl) that even though an estimated one million cubic miles of the earth's groundwater is located within about half a mile of the surface (there are about one billion gallons in a cubic mile), only a small amount of this reservoir of underground water can be tapped and made available for human use through wells and springs. Furthermore, in *Where Is Earth's Water Located?* (August 28, 2006, http://ga.water.usgs.gov/edu/earthwherewater.html), the USGS notes that the total amount of fresh groundwater on Earth at all depths is estimated at about 2.5 million cubic miles.

HOW GROUNDWATER OCCURS

Groundwater is not in underground lakes, nor is it water flowing in underground rivers. It is simply water that fills pores or cracks in subsurface rocks. When rain falls or snow melts on the surface of the ground, some water may run off into lower land areas or lakes and streams. What is left may be absorbed by the soil, seep into deeper layers of soil and rock, or evaporate into the atmosphere.

Below the topsoil—the rich upper layer of soil in which plants have most of their roots—is an area called the unsaturated zone. In times of adequate rainfall the small spaces between rocks and grains of soil in the unsaturated zone contain at least some water, whereas the larger spaces contain mostly air. After a major rain, however, all the open spaces may fill with water temporarily. During a drought, the area may become drained and almost completely dry, although a certain amount of water is held in the soil and rocks by molecular attraction.

Lying beneath the unsaturated zone is the saturated zone. The water table is the level at which the unsaturated zone and the saturated zone meet. (See Figure 4.1.) Water drains through the unsaturated zone to the saturated zone. The saturated zone is full of water—all the spaces between soil and rocks, and within the rocks themselves, contain water. Water from streams, lakes, wetlands, and other water bodies may seep into the saturated zone. Streams are commonly a significant source of recharge to groundwater downstream from mountain fronts and steep hillsides in arid and semiarid areas, and in areas underlaid by limestone and other porous rock.

The water table is not fixed, but may rise or fall, depending on water availability. In areas where the climate is fairly consistent, the level of the water table may vary little; in areas subject to extreme flooding and drought, it may rise and fall substantially.

Groundwater Flow

Water is always in motion. Groundwater generally moves from recharge areas, where water enters the ground, to discharge areas, where it exits from the ground into a wetland, river, lake, or ocean. Transpiration by plants whose roots extend to a point near the water table is another form of discharge. The path of groundwater movement may be short and simple or incredibly complex, depending on the geology of the area through which the water passes. The complexity of the path also determines the length of time a molecule of water remains in the ground between recharge and discharge points. (See Figure 4.2.)

FIGURE 4.1

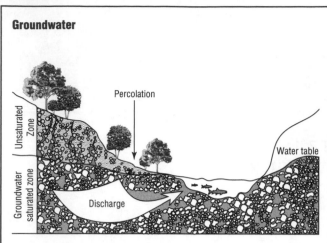

Groundwater

Percolation

Water table

Unsaturated Zone

Groundwater saturated zone

Discharge

SOURCE: "Figure 1-1. Ground Water," in *National Water Quality Inventory: 1998 Report to Congress*, U.S. Environmental Protection Agency, Office of Water, June 2000, http://www.epa.gov/305b/98report/chap1.pdf (accessed January 8, 2007)

FIGURE 4.2

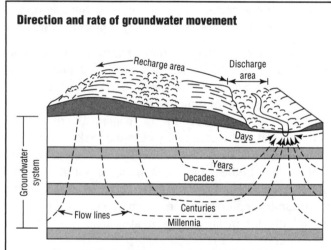

Direction and rate of groundwater movement

Recharge area

Discharge area

Days

Years

Decades

Groundwater system

Centuries

Flow lines

Millennia

SOURCE: Roger M. Waller, "Direction and Rate of Ground-Water Movement," in *Ground Water and the Rural Homeowner*, U.S. Department of the Interior, U.S. Geological Survey, August 19, 2005, http://pubs.usgs.gov/gip/gw_ruralhomeowner/ (accessed January 4, 2007)

The velocities of groundwater flow generally are low and are orders of magnitude less than the velocities of stream flow. Groundwater movement normally occurs as slow seepage through the spaces between particles of unconsolidated material or through networks of fractures and openings in consolidated rocks. A velocity of one foot per day or more is a high rate of movement in groundwater. Groundwater velocities can be as low as one foot per decade or one foot per century. By contrast, stream flows are generally measured in feet per second. A velocity of one foot per second is about sixteen miles per day. The low velocities of groundwater flow can have

important implications, particularly in relation to the movement of contaminants.

The age of water (time since recharge) varies in different parts of groundwater flow systems. Groundwater gets steadily older along a particular flow path from an area of recharge to an area of discharge. In shallow, local-scale flow systems, groundwater age at areas of discharge can vary from less than a day to a few hundred years. (See Figure 4.2.) In deep, regional flow systems with long flow paths, groundwater age may reach thousands or tens of thousands of years.

AQUIFERS

An aquifer is a saturated zone that contains enough water to yield significant amounts of water when a well is dug. The zone is actually a path of porous or permeable material through which substantial quantities of water flow relatively easily. The word *aquifer* comes from the Latin *aqua* (water) and *ferre* (to bear or carry). An aquifer can be a layer of gravel or sand, a layer of sandstone or cavernous limestone, a rubble zone between lava flows, or even a large body of massive rock, such as fractured granite. An aquifer may lie above, below, or in between confining beds that are layers of hard, nonporous material (e.g., clay or solid granite).

There are two types of aquifers: unconfined and confined (artesian). In an unconfined or water table aquifer, precipitation filters down from the land's surface until it hits an impervious layer of rock or clay. The water then accumulates and forms a zone of saturation. Because runoff water can easily seep down to the water table, an unconfined aquifer is susceptible to contamination.

In a confined aquifer the confining beds act more or less like underground boundaries, discouraging water from entering or leaving the aquifer, so that the water is forced to continue its slow movement to its discharge point. Water from precipitation enters the aquifer through a recharge area, where the soil lets the water percolate down to the level of the aquifer. The ability of an aquifer to recharge is dependent on various factors, such as the ease with which water is able to move down through the geological formations (permeability) and the size of the spaces between the rock particles (porosity). Figure 4.3 illustrates natural and artificial aquifer recharge.

Usually, the permeability and porosity of rocks decreases as their depth below the surface increases. How much water can be removed from an aquifer depends on the type of rock. A dense granite, for example, will supply almost no water to a well even though the water is near the surface. A porous sandstone, however, thousands of feet below the surface can yield hundreds of gallons of water per minute. Porous rocks that are capable of supplying freshwater have been found at depths of

FIGURE 4.3

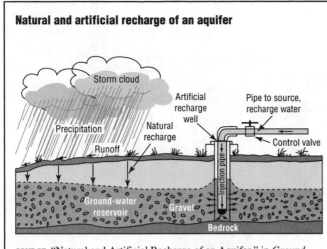

Natural and artificial recharge of an aquifer

SOURCE: "Natural and Artificial Recharge of an Aquifer," in *Ground Water*, U.S. Department of the Interior, U.S. Geological Survey, 1999, http://capp.water.usgs.gov/GIP/gw_gip/gwgip.pdf (accessed January 4, 2007)

more than six thousand feet below the surface. Saline (salty) water has been discovered in aquifers that lie more than thirty thousand feet underground.

Aquifers vary from a few feet thick to tens or hundreds of feet thick. They can be located just below the earth's surface or thousands of feet beneath it. An aquifer can cover a few acres of land or many thousands of square miles. Furthermore, any one aquifer may be a part of a large system of aquifers that feed into each other.

Ogallala Aquifer

The Ogallala or High Plains Aquifer in the United States is one of the world's largest aquifers and is the largest in North America. According to the High Plains Aquifer Information Network (2007, http://www.hiplain .org/states/index.cfm?state=9&c=1&sc=84), the aquifer stretches from southern South Dakota to the Texas panhandle and covers 174,000 square miles. (See Figure 4.4.) The Ogallala's average thickness ranges from three hundred feet to more than twelve hundred feet in Nebraska. For more than fifty years, the Ogallala has supplied most of the water for irrigation and drinking to the Great Plains states and yields about 30% of the nation's groundwater for irrigation. Robert B. Jackson et al. indicate in "Water in a Changing World" (*Issues in Ecology*, Spring 2001) that since 1940 approximately 200,000 wells tap into this vast aquifer. Furthermore, the USGS (February 16, 2007, http://webserver.cr.usgs.gov/ nawqa/hpgw/HPGW_home.html) notes that in 2000 these wells extracted about 315 million gallons of water per day for 1.9 million people. The U.S. Environmental Protection Agency (EPA) designates the Ogallala aquifer a sole-source aquifer. This means that at least 50% of the population in the area depends on the aquifer for its water supply.

Paul D. Ryder reports in *Ground Water Atlas of the United States, Oklahoma, Texas* (1996, http://capp.water. usgs.gov/gwa/ch_e/index.html) that because the Ogallala Aquifer is being pumped far in excess of recharge, the USGS and the Texas Department of Water Resources project an increasing shortage of Ogallala Aquifer water for future irrigation needs. The High Plains Aquifer Information Network agrees that pumping for irrigation from the aquifer has resulted in substantial declines in some of its parts. Projections by the USGS and the Texas Department of Water Resources suggest that the irrigated acreage in the High Plains of Texas (69% of irrigated Texas cropland) will be reduced to half of its present acreage by 2030 unless an effective water conservation plan is implemented.

In 1999 the Texas oil tycoon Boone Pickens formed Mesa Water, Inc., to market water from part of the Ogallala Aquifer to large Texas cities for municipal use. About one hundred landowners and two hundred thousand acres of land in the Texas panhandle were affected by Pickens's plan. The project sparked controversy in Texas, where the hundred-year-old rule of capture was still in effect. The rule of capture, which was at one time standard doctrine in much of the United States, states that the owner of land that sits on an underground water source can pump out unlimited amounts of water regardless of the impact on surrounding property owners.

Concerned that the already rapidly draining water source would become depleted even further, residents of the surrounding panhandle area protested, arguing that their essential source of water should not be pumped hundreds of miles away. Many Texans began campaigning for tighter government regulations on water rights. However, comprehensive water legislation that would have modified the rule of capture failed to reach a vote in the Texas state senate in May 2005. The following month, the article "Private Water Group Preparing to Pump the Ogallala Aquifer and Sell Groundwater to Far-Away Texas Cities" (*Western Water Law and Policy Reporter*, June 2005) announced that Mesa Water would begin building a system of wells and pipelines to sell the water to major urban centers throughout Texas.

According to the article "Deal Beefs up Authority's Water Rights" (*Amarillo Daily News*, March 7, 2006), the Canadian River Municipal Water Authority (CRMWA) announced in March 2006 interest in all of Mesa's two hundred thousand acres of water rights. The CRMWA was created over fifty years ago by the Texas legislature to provide a source of municipal and industrial water for its eleven member cities in the Texas panhandle and South Plains via a 322-mile aqueduct system. (The Canadian River is the largest tributary of the Arkansas River and runs through this part of Texas.)

FIGURE 4.4

High Plains aquifer system

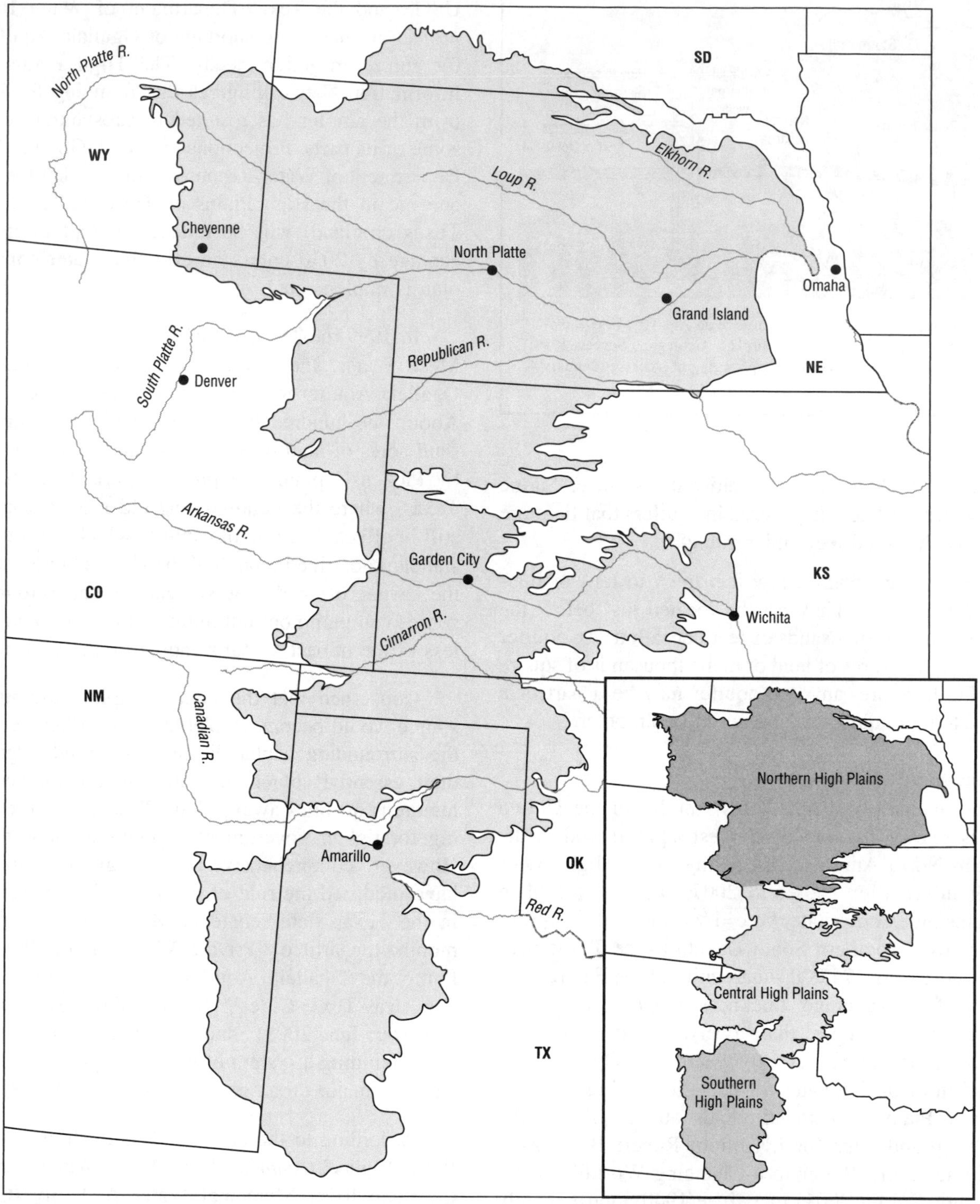

SOURCE: "High Plains Regional Ground-Water Study," in *National Water-Quality Assessment (NAWQA) Program—High Plains Regional Ground-Water (HPGW) Study*, U.S. Department of the Interior, U.S. Geological Survey, 2006, http://co.water.usgs.gov/nawqa/hpgw/HPGW_home.html (accessed January 5, 2007)

SPRINGS

A spring is a natural discharge of water at the earth's surface from a saturated zone that has been filled to overflowing. Springs are classified either according to the amount of water they produce or according to the temperature of the water (hot, warm, or cold). Giant

Springs in Great Falls, Montana, is the largest freshwater spring in the United States and is the source of water for the Missouri and Roe rivers. According to Travel Montana (2007, http://montanakids.com/db_engine/presentations/presentation.asp?pid=48&sub=Giant+Springs), the springs remove 338 million gallons per day from underground reserves. Furthermore, the water remains a constant temperature of 54°F, and has been carbon-dated to be about three thousand years old.

Thermal springs have water that is warm or, in some places, hot. They are fed by groundwater that is heated by contact with hot rocks deep below the surface. In some areas water can descend slowly to deep levels, getting warmer the farther down it goes. If it rises faster than it descended, it does not have time to cool off before it emerges on the surface. Well-known thermal springs are the Warm Springs in Georgia and the Hot Springs in Arkansas. Geysers are thermal springs that erupt periodically. Old Faithful in Yellowstone National Park is perhaps the most famous and spectacular geyser in the world. It erupts at intervals of thirty to ninety minutes. The park maintains a Webcam (http://www.nps.gov/archive/yell/oldfaithfulcam.htm) so that virtual visitors can see Old Faithful erupt even if they cannot see it in person.

NATURAL CHARACTERISTICS OF GROUNDWATER

As groundwater travels its course from recharge to discharge area, it undergoes chemical and physical changes as it mixes with other groundwater and reacts with the minerals in the sand or rocks through which it flows. These interactions can greatly affect water quality and its suitability or unsuitability for a particular use.

Minerals

Water is a natural solvent capable of dissolving many other substances. Spring waters may contain dissolved minerals and gases that give them subtle flavors. Without minerals and gases, water tastes flat. The most common dissolved mineral substances are calcium, magnesium, sodium, potassium, chloride, sulfate, and bicarbonate. However, water is not considered desirable for drinking if it contains more than one thousand milligrams per liter (mg/l) of dissolved minerals. In areas where less-mineralized water is not available, water with a few thousand mg/l of dissolved minerals is used routinely, although it is classified as saline.

Some well and spring waters contain such high levels of dissolved minerals that they cannot be tolerated by humans, plants, or animals. In high concentrations, certain minerals can be especially harmful. A large quantity of sodium in drinking water is unhealthy for people with heart disease. Boron, a mineral that is good for some plants in small amounts, is toxic to other plants in only

slightly elevated concentrations. Such highly mineralized groundwater usually lies deep below the surface and has limited uses.

Water Hardness

Water that contains a lot of calcium and magnesium is said to be hard. The hardness of water can be expressed in terms of the amount of calcium carbonate (the principal constituent of limestone) or equivalent minerals that remain when the water is evaporated. Water is considered soft when it contains 0 to 60 mg/l of hardness constituents, moderately hard from 61 to 120 mg/l, hard between 121 and 180 mg/l, and very hard if more than 180 mg/l.

Very hard water is not desirable for many domestic uses and leaves a scaly deposit on the insides of pipes, boilers, and tanks. Hard water can be made soft at a fairly reasonable cost, although it is not always desirable to remove all the minerals from drinking water because some are beneficial to health. Extremely soft water can corrode metals but is suitable for doing laundry, dishwashing, and bathing. Whenever possible, most communities seek a balance between hard and soft water in their municipal water systems.

CURRENT GROUNDWATER USE
Human Needs

The nation's use of groundwater grew dramatically in the last several decades of the twentieth century. Susan S. Hutson et al. report in *Estimated Use of Water in the United States in 2000* (2004, http://pubs.usgs.gov/circ/2004/circ1268/pdf/circular1268.pdf) that the rate of withdrawal was thirty-four billion gallons per day (Bgal/d) of fresh groundwater in 1950. It increased to 83 Bgal/d in 1980 and dropped off a bit over the next several years before reaching a new high of 83.3 Bgal/d in 2000.

In "Ground Water Use in the United States" (March 27, 2006, http://ga.water.usgs.gov/edu/wugw.html), the USGS estimates that approximately 26% of the freshwater used in the United States in 2000 was groundwater. (The rest was surface water.) Figure 4.5 shows the percentage of groundwater allocated to various uses in the United States in 2000. Nearly two-thirds (63%) of all groundwater was used for irrigation. One-fifth (20%) was used for public uses such as drinking, bathing, and cooking. The remaining 17% was used for industry, mining, domestic use (self-supplied water via wells), livestock watering, thermoelectric power plants, and commercial purposes.

Figure 4.6 shows the estimated percent of the population using groundwater as drinking water. In many states—including Idaho, Minnesota, Nebraska, New Mexico, Mississippi, and Florida—drinking water is obtained almost exclusively from groundwater sources. According John S. Zogorski et al., in *The Quality of Our Nation's Waters:*

FIGURE 4.5

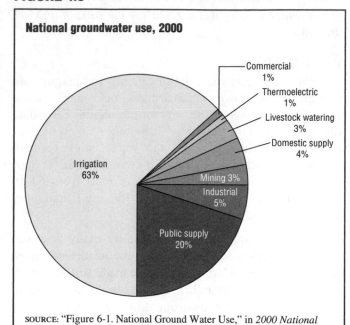

National groundwater use, 2000

- Commercial 1%
- Thermoelectric 1%
- Livestock watering 3%
- Domestic supply 4%
- Mining 3%
- Industrial 5%
- Public supply 20%
- Irrigation 63%

SOURCE: "Figure 6-1. National Ground Water Use," in *2000 National Water Quality Inventory*, U.S. Environmental Protection Agency, Office of Water, August 2002, http://www.epa.gov/305b/2000report/chp6.pdf (accessed January 8, 2007)

Volatile Organic Compounds in the Nation's Ground Water and Drinking-Water Supply Wells (2006, http://pubs.usgs.gov/circ/circ1292/pdf/circular1292.pdf), overall about 50% of U.S. residents use groundwater as their drinking water source. Rural residents rely heavily on groundwater for this purpose.

Ecological Needs

Historically, groundwater and surface water have been managed as separate resources. Since the 1970s, however, there has been a growing awareness that these two sources are inseparably linked. Groundwater seeps into rivers, streams, lakes, and other water bodies and breaks the surface as springs. In some parts of the United States, especially in arid regions, aquifers contribute a large portion of the water found in rivers and streams. Figure 4.7 shows groundwater contributions to surface water in ten regions across the country.

Groundwater recharge of surface water is particularly important during dry periods. Reductions in surface water can have adverse effects on the ecology of a watershed, stressing fish populations and their food supply, wetlands, and the plants and animals living along the banks of rivers and streams. Groundwater depletion in some areas has resulted in the death of aquatic and semiaquatic species that depended on groundwater flow to surface water streams.

Overpumping

Pumping groundwater from a well always causes a decline in groundwater levels at and near the well, and it always causes a diversion of groundwater that was moving slowly to its natural, possibly distant, area of discharge. Pumping a single well typically has only a local effect on the groundwater flow system. Pumping many wells (sometimes hundreds or thousands of wells) in large areas can have significant regional effects on groundwater systems.

If a groundwater system is not overused, the rate of groundwater recharge and discharge balance one another. However, when the rate of withdrawal exceeds the rate at which the groundwater source is recharged, the result is the lowering of groundwater to levels that may impair the resource.

Overpumping groundwater can have many different effects, including:

- Neighboring wells can dry up, requiring construction of new, deeper wells or significant changes to existing wells.

- Aquifer materials can compact, causing the land above the aquifer to sink and leaving gaping holes in the land that cause damage to buildings, roads, canals, pipelines, and other infrastructure.

- Aquifer capacity may be permanently lost because of compaction of aquifer materials, resulting in higher pumping costs and a decrease in well yields.

- Changes in the volume and direction of groundwater flow can induce the flow of saltwater and water of lower quality into a well.

- Wetlands can dry up and cause adverse effects on ecological systems that are dependent on groundwater discharge.

According to the USGS, large withdrawals of groundwater have altered the flow systems and geological and chemical conditions of some of the major aquifers in the United States. Declining groundwater levels can change the location and size of recharge areas and reduce discharge rates. Some aquifers in the West have suffered major losses in aquifer storage because of overpumping.

VULNERABLE RESOURCE—GROUNDWATER QUALITY

Until the mid-twentieth century people believed that soil provided a barrier or protective filter that neutralized the downward migration of contaminants from the land surface and prevented water resources from becoming contaminated. The discovery of pesticides and contaminants in groundwater, however, demonstrated that human activities do influence groundwater quality and that the soil may not be as effective a filter as once thought.

FIGURE 4.6

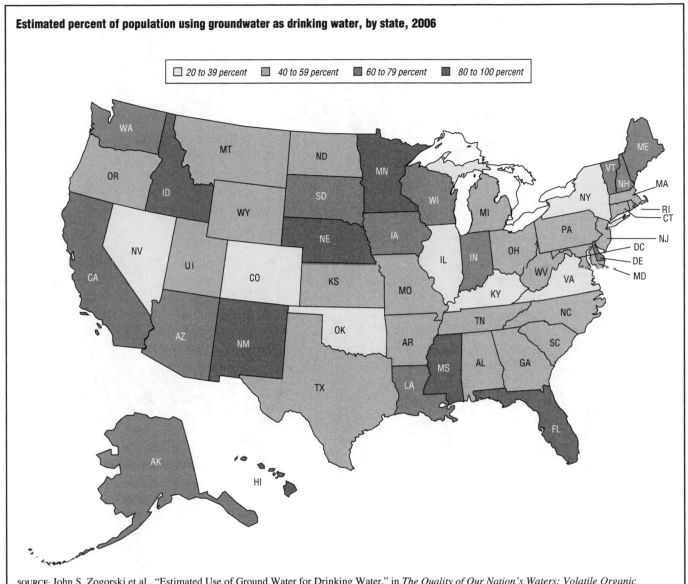

Estimated percent of population using groundwater as drinking water, by state, 2006

| | 20 to 39 percent | | 40 to 59 percent | | 60 to 79 percent | | 80 to 100 percent |

SOURCE: John S. Zogorski et al., "Estimated Use of Ground Water for Drinking Water," in *The Quality of Our Nation's Waters: Volatile Organic Compounds in the Nation's Ground Water and Drinking-Water Supply Wells*, U.S. Department of the Interior, U.S. Geological Survey, 2006, http://pubs .usgs.gov/circ/circ1292/pdf/circular1292.pdf (accessed January 24, 2007)

The potential for a contaminant to affect groundwater quality is dependent on its ability to migrate through the overlying soils to the groundwater resource. Figure 4.8 shows sources of groundwater contamination. Contamination can occur as a relatively well-defined localized plume coming from a specific source. It can also occur as a generalized deterioration over a large area because of diffuse nonpoint sources such as fertilizer and pesticide applications.

Once groundwater contamination was recognized, researchers needed to determine which waters were contaminated, the severity of contamination, and what should be done about the contamination. Many government and private organizations began working to find the answers, but it was not an easy task.

The quality of the most available groundwater in the United States is believed to be good, according to the EPA's *Safe Drinking Water Act, Section 1429 Groundwater Report to Congress* (October 1999, http:// www.gwpc.org/e-Library/Documents/GW_Report_to_ Congress.pdf) and *2000 National Water Quality Inventory* (August 2002, http://www.epa.gov/305b/2000report/). The worst groundwater contamination is generally in the areas where use is heaviest—towns and cities, industrial complexes, and agricultural regions, such as California's Central Valley.

Recognizing the need to protect valuable and vulnerable groundwater sources, the states have begun to implement comprehensive groundwater protection

FIGURE 4.7

Groundwater contributions to surface water in ten hydrologic regions around the country

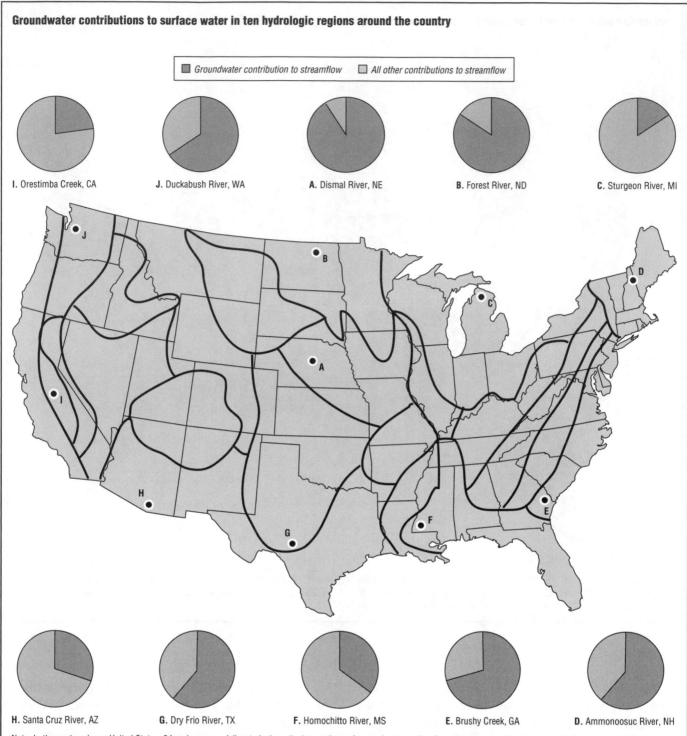

■ Groundwater contribution to streamflow   ■ All other contributions to streamflow

I. Orestimba Creek, CA

J. Duckabush River, WA

A. Dismal River, NE

B. Forest River, ND

C. Sturgeon River, MI

H. Santa Cruz River, AZ

G. Dry Frio River, TX

F. Homochitto River, MS

E. Brushy Creek, GA

D. Ammonoosuc River, NH

Note: In the conterminous United States, 24 regions were delineated where the interactions of groundwater and surface water are considered to have similar characteristics (i.e., groundwater accounts for a similar proportion of surface water discharge within each region). The estimated groundwater contribution to stream flow is shown for specific streams in 10 of the regions.

SOURCE: "Exhibit 2-5. Ground Water Contributions to Surface Water in 10 Hydrologic Regions Around the Country," in *Safe Drinking Water Act, Section 1429 Ground Water Report to Congress*, U.S. Environmental Protection Agency, Office of Water, October 1999, http://www.gwpc.org/e-Library/ Documents/GW_Report_to_Congress.pdf (accessed January 8, 2007)

programs. In addition, in October 2006 the EPA finalized the Ground Water Rule (October 25, 2006, http://www.epa .gov/safewater/disinfection/gwr/basicinformation.html #six), which requires states to conduct sanitary surveys of water provided by public utilities from underground sources, take corrective action if contamination is found, and ensure that disinfection of drinking water is effective.

FIGURE 4.8

Sources of groundwater contamination

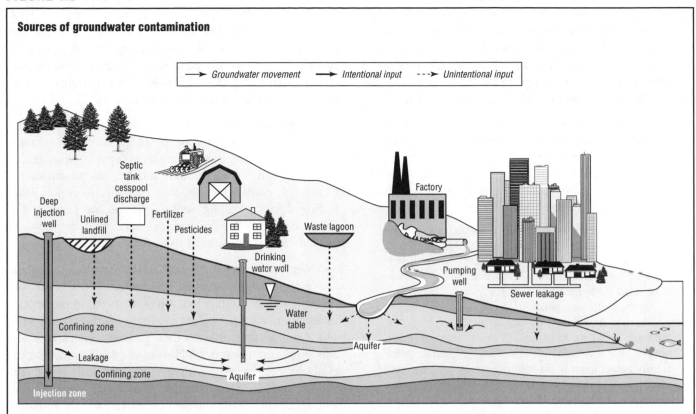

SOURCE: "Figure 6-2. Sources of Ground Water Contamination," in *2000 National Water Quality Inventory*, U.S. Environmental Protection Agency, Office of Water, August 2002, http://www.epa.gov/305b/2000report/chp6.pdf (accessed January 5, 2007)

FACTORS AFFECTING GROUNDWATER CONTAMINATION

All pollutants do not cause the same rate of contamination for the same amount of pollutant. Groundwater is affected by many of the following factors:

- The distance between the land surface where pollution occurs and the depth of the water table. The greater the distance, the greater the chance that the pollutant will biodegrade or react with soil minerals.

- The mineral composition of the soil and rocks in the unsaturated zone. Heavy soil and organic materials lessen the potential for contamination.

- The presence or absence of biodegrading microbes in the soil.

- The amount of rainfall. Less rainfall results in less water entering the saturated zone and, therefore, lower quantities of contaminants.

- The evapotranspiration rate. (This is the rate at which water is discharged into the atmosphere as a result of evaporation from the soil, surface water, and plants.) High rates reduce the amount of contaminated water reaching the saturated zone.

GROUNDWATER CONTAMINATION

Major Types of Groundwater Contaminants

The EPA reports in the *National Water Quality Inventory: 1998 Report to Congress* (June 2000, http://www.epa.gov/305b/98report/) that thirty-one of the thirty-seven reporting states identified the types of contaminants they found in groundwater. The states said that nitrates, metals, volatile and semivolatile organic compounds, and pesticides were the pollutants found most often. Following the publication of the *2000 National Water Quality Inventory*, the EPA entered a transition period in the gathering and analysis of water quality data in nationally consistent, statistically valid assessment reports. Its new reporting schedule is described in "Schedule for Statistically Valid Surveys of the Nation's Waters" (December 5, 2005, http://www.epa.gov/owow/monitoring/guide.pdf). The USGS conducts the National Water Quality Assessment Program and publishes a series of reports on water quality issues of regional and national concern.

In *Factors Affecting Occurrence and Distribution of Selected Contaminants in Ground Water from Selected Areas in the Piedmont Aquifer System, Eastern United States, 1993–2003* (2006, http://pubs.usgs.gov/sir/2006/5104/pdf/sir2006-5104.pdf), Bruce D. Lindsey et al.

discuss the Piedmont Aquifer System (PAS), which is a fingerlike area extending from Pennsylvania and New Jersey in the north to Georgia and Alabama in the south. It is a major aquifer in the eastern United States that follows the eastern foothills of the Appalachian Mountains. Lindsey et al. sampled wells and springs in the PAS as part of the USGS's National Water Quality Assessment Program.

In general, Lindsey et al. provide a positive report concerning groundwater contaminants in this aquifer. In the news release about the report, "Ground Water Meets Most Federal Standards in Major Eastern U.S. Aquifer" (December 20, 2006, http://www.usgs.gov/newsroom/article.asp?ID=1593), the USGA states:

> Many chemicals were detected in ground water from selected areas of the Piedmont Aquifer System (PAS), but concentrations of those chemicals were below drinking-water standards in most cases. ... The findings in the PAS, based on samples from 255 wells and 19 springs, do not generally imply present human-health risk; however, they are an early warning that land-use activities have an effect on regional water quality. For example, concentrations of nitrate were significantly higher in ground water underlying agricultural land use than in ground water underlying undeveloped or urban land. Herbicides were detected more frequently in agricultural wells, whereas insecticides, VOCs [volatile organic compounds], chloroform, and MTBE [a fuel component in gasoline] were more frequently detected in urban wells.
>
> Findings also show that rock settings can have a great effect on ground-water quality, particularly for radon, a natural product from the radioactive decay of uranium.

A list of drinking water contaminants, their sources, and their health effects is shown in Table 4.1. Some of the more common groundwater contaminants are described in the following sections.

ARSENIC. Arsenic is a naturally occurring element in rocks and soils and is the twentieth most common element in the earth's crust. The presence of arsenic in groundwater is largely the result of minerals dissolving from naturally weathered rocks and soils. The USGS reports that the nation's groundwater typically contains less than one or two parts per billion (ppb) of arsenic. One ppb is equal to approximately one teaspoon of powdered arsenic in two Olympic-sized swimming pools. Moderate to high arsenic levels do occur in some areas throughout the nation due to geology, geochemistry, and climate. Elevated arsenic concentrations in groundwater are commonly found in the West and in parts ·of the Midwest and the Northeast.

Arsenic research shows that humans need arsenic as a trace element in their diet to survive. Too much arsenic, however, can be harmful. Paolo Boffetta and Fredrik Nyberg indicate in "Contribution of Environmental Factors to Cancer Risk" (*British Medical Bulletin*, 2003) that

prolonged exposure to arsenic can contribute to skin, bladder, and other cancers. In January 2001 the EPA proposed lowering the current maximum contaminant level (MCL) for arsenic in drinking water from fifty ppb to ten ppb. The effective date of the new arsenic rule was February 22, 2002. Public water systems had to comply with the ten-ppb level by January 23, 2006.

NITRATES. Many scientists and geologists consider nitrates to be the most widespread groundwater contaminant. Nitrates are simply another form of nitrogen, a plant nutrient. Nitrogen and nitrates, as discussed in Chapter 3, enter bodies of water usually as runoff from fertilized land, leaking septic systems, or sewage discharges. Generally, a level of three ppb or more in groundwater is considered indicative of human impact.

Nitrate contamination occurs most frequently in shallow groundwater (less than one hundred feet below the surface) and in aquifers that allow the rapid movement of water. Regional differences in nitrate levels are related to soil drainage properties, other geologic characteristics, and agricultural practices. Nitrate in groundwater is generally highest in areas with well-drained soils and intensive cultivation of row crops, particularly corn, cotton, and vegetables. Low nitrate concentrations are found in areas of poorly drained soil and where pasture and woodland are intermixed with cropland. Crop fertilization is the most important agricultural practice for introducing nitrogen into groundwater. The primary source of nitrates is fertilizers used in agriculture and, in some areas, feedlot operations.

Nitrates are important because they affect both human and ecological health. They can cause a public health risk to infants and young livestock. In some areas of the country substantial amounts of nitrates in surface water are contributed by groundwater sources.

PESTICIDES. Robert J. Gilliom et al. report in *The Quality of Our Nation's Water: Pesticides in the Nation's Streams and Ground Water, 1992–2001* (February 15, 2007, http://pubs.usgs.gov/circ/2005/1291/pdf/circ1291.pdf) that pesticides are found less frequently in groundwater than in surface water. Nonetheless, pesticides and their broken-down products are found frequently in shallow groundwater, especially in residential and agricultural areas. However, the pesticides are rarely found in concentrations exceeding water quality benchmarks for human health.

VOCs. Volatile organic compounds (VOCs) are those that contain the element carbon and tend to evaporate more quickly than water. Examples of substances that contain a variety of volatile organic compounds are gasoline, diesel fuel, paint, glue, spot removers, and cleaning solutions. VOCs are used extensively in industry to manufacture such products as cars, electronics, computers,

TABLE 4.1

Drinking water contaminants, their sources, and potential health effects, 2003

Type	Contaminant	MCL or TT[a] (mg/L)[b]	Potential health effects from exposure above the MCL	Common sources of contaminant in drinking water	Public health goal
OC	Acrylamide	TT[h]	Nervous system or blood problems; increased risk of cancer	Added to water during sewage/wastewater treatment	zero
OC	Alachlor	0.002	Eye, liver, kidney or spleen problems; anemia; increased risk of cancer	Runoff from herbicide used on row crops	zero
R	Alpha particles	15 picocuries per liter (pCi/L)	Increased risk of cancer	Erosion of natural deposits of certain minerals that are radioactive and may emit a form of radiation known as alpha radiation	zero
IOC	Antimony	0.006	Increase in blood cholesterol; decrease in blood sugar	Discharge from petroleum refineries; fire retardants; ceramics; electronics; solder	0.006
IOC	Arsenic	0.010 as of 1/23/06	Skin damage or problems with circulatory systems, and may have increased risk of getting cancer	Erosion of natural deposits; runoff from orchards, runoff from glass & electronics production wastes	C
IOC	Asbestos (fibers >10 micrometers)	7 million fibers per liter (MFL)	Increased risk of developing benign intestinal polyps	Decay of asbestos cement in water mains; erosion of natural deposits	7 MFL
OC	Atrazine	0.003	Cardiovascular system or reproductive problems	Runoff from herbicide used on row crops	0.003
IOC	Barium	2	Increase in blood pressure	Discharge of drilling wastes; discharge from metal refineries; erosion of natural deposits	2
OC	Benzene	0.005	Anemia; decrease in blood platelets; increased risk of cancer	Discharge from factories; leaching from gas storage tanks and landfills	zero
OC	Benzo(a)pyrene (PAHs)	0.0002	Reproductive difficulties; increased risk of cancer	Leaching from linings of water storage tanks and distribution lines	zero
IOC	Beryllium	0.004	Intestinal lesions	Discharge from metal refineries and coal-burning factories; discharge from electrical, aerospace, and defense industries	0.004
R	Beta particles and photon emitters	4 millirems per year	Increased risk of cancer	Decay of natural and man-made deposits of certain minerals that are radioactive and may emit forms of radiation known as photons and beta radiation	zero
DBP	Bromate	0.010	Increased risk of cancer	By-product of drinking water disinfection	zero
IOC	Cadmium	0.005	Kidney damage	Corrosion of galvanized pipes; erosion of natural deposits; discharge from metal refineries; runoff from waste batteries and paints	0.005
OC	Carbofuran	0.04	Problems with blood, nervous system, or reproductive system	Leaching of soil fumigant used on rice and alfalfa	0.04
OC	Carbon tetrachloride	0.005	Liver problems; increased risk of cancer	Discharge from chemical plants and other industrial activities	zero
D	Chloramines (as Cl$_2$)	MRDL=4.0[a]	Eye/nose irritation; stomach discomfort, anemia	Water additive used to control microbes	MRDLG=4[a]
OC	Chlordane	0.002	Liver or nervous system problems; increased risk of cancer	Residue of banned termiticide	zero
D	Chlorine (as Cl$_2$)	MRDL=4.0[a]	Eye/nose irritation; stomach discomfort	Water additive used to control microbes	MRDLG=4[a]
D	Chlorine dioxide (as ClO$_2$)	MRDL=0.8[a]	Anemia; infants & young children: nervous system effects	Water additive used to control microbes	MRDLG=0.8[a]
DBP	Chlorite	1.0	Anemia; infants & young children: nervous system effects	By-product of drinking water disinfection	0.8
OC	Chlorobenzene	0.1	Liver or kidney problems	Discharge from chemical and agricultural chemical factories	0.1
IOC	Chromium (total)	0.1	Allergic dermatitis	Discharge from steel and pulp mills; erosion of natural deposits	0.1
IOC	Copper	TT[g]; action level=1.3	Short-term exposure: gastrointestinal distress. Long-term exposure: liver or kidney damage. People with Wilson's Disease should consult their personal doctor if the amount of copper in their water exceeds the action level	Corrosion of household plumbing systems; erosion of natural deposits	1.3
M	Cryptosporidium	TT[c]	Gastrointestinal illness (e.g., diarrhea, vomiting, cramps)	Human and animal fecal waste	zero
IOC	Cyanide (as free cyanide)	0.2	Nerve damage or thyroid problems	Discharge from steel/metal factories; discharge from plastic and fertilizer factories	
OC	2,4-D	0.07	Kidney, liver, or adrenal gland problems	Runoff from herbicide used on row crops	
OC	Dalapon	0.2	Minor kidney changes	Runoff from herbicide used on rights of way	

adhesives, dyes, and plastics; they are also used in dry cleaning and refrigeration.

VOCs can cause cancer, have adverse effects on various body organs and systems, and affect the brain, ears, eyes, skin, and throat. Groundwater contamination can occur from landfills, hazardous waste facilities, and septic systems into which VOCs have been discarded, or from sources such as leaking underground storage tanks. They can be released into the environment from industry, enter the atmosphere, and fall to the ground as atmospheric

TABLE 4.1

Drinking water contaminants, their sources, and potential health effects, 2003 [CONTINUED]

Type	Contaminant	MCL or TT[a] (mg/L)[b]	Potential health effects from exposure above the MCL	Common sources of contaminant in drinking water	Public health goal
OC	1,2-Dibromo-3-chloropropane (DBCP)	0.0002	Reproductive difficulties; increased risk of cancer	Runoff/leaching from soil fumigant used on soybeans, cotton, pineapples, and orchards	zero
OC	o-Dichlorobenzene	0.6	Liver, kidney, or circulatory system problems	Discharge from industrial chemical factories	0.6
OC	p-Dichlorobenzene	0.075	Anemia; liver, kidney or spleen damage; changes in blood	Discharge from industrial chemical factories	0.075
OC	1,2-Dichloroethane	0.005	Increased risk of cancer	Discharge from industrial chemical factories	zero
OC	1,1-Dichloroethylene	0.007	Liver problems	Discharge from industrial chemical factories	0.007
OC	cis-1,2-Dichloroethylene	0.07	Liver problems	Discharge from industrial chemical factories	0.07
OC	trans-1,2-Dichloroethylene	0.1	Liver problems	Discharge from industrial chemical factories	0.1
OC	Dichloromethane	0.005	Liver problems; increased risk of cancer	Discharge from drug and chemical factories	zero
OC	1,2-Dichloropropane	0.005	Increased risk of cancer	Discharge from industrial chemical factories	zero
OC	Di(2-ethylhexyl) adipate	0.4	Weight loss, liver problems, or possible reproductive difficulties	Discharge from chemical factories	0.4
OC	Di(2-ethylhexyl) phthalate	0.006	Reproductive difficulties; liver problems; increased risk of cancer	Discharge from rubber and chemical factories	zero
OC	Dinoseb	0.007	Reproductive difficulties	Runoff from herbicide used on soybeans and vegetables	0.007
OC	Dioxin (2,3,7,8-TCDD)	0.00000003	Reproductive difficulties; increased risk of cancer	Emissions from waste incineration and other combustion; discharge from chemical factories	zero
OC	Diquat	0.02	Cataracts	Runoff from herbicide use	0.02
OC	Endothall	0.1	Stomach and intestinal problems	Runoff from herbicide use	0.1
OC	Endrin	0.002	Liver problems	Residue of banned insecticide	0.002
OC	Epichlorohydrin	TT[h]	Increased cancer risk, and over a long period of time, stomach problems	Discharge from industrial chemical factories; an impurity of some water treatment chemicals	zero
OC	Ethylbenzene	0.7	Liver or kidneys problems	Discharge from petroleum refineries	0.7
OC	Ethylene dibromide	0.00005	Problems with liver, stomach, reproductive system, or kidneys; increased risk of cancer	Discharge from petroleum refineries	zero
IOC	Fluoride	4.0	Bone disease (pain and tenderness of the bones); children may get mottled teeth	Water additive which promotes strong teeth; erosion of natural deposits; discharge from fertilizer and aluminum factories	4.0
M	Giardia lamblia	TT[c]	Gastrointestinal illness (e.g., diarrhea, vomiting, cramps)	Human and animal fecal waste	zero
OC	Glyphosate	0.7	Kidney problems; reproductive difficulties	Runoff from herbicide use	0.7
DBP	Haloacetic acids (HAA5)	0.060	Increased risk of cancer	By-product of drinking water disinfection	n/a[f]
OC	Heptachlor	0.0004	Liver damage; increased risk of cancer	Residue of banned termiticide	zero
OC	Heptachlor epoxide	0.0002	Liver damage; increased risk of cancer	Breakdown of heptachlor	zero
M	Heterotrophic plate count (HPC)	TT[c]	HPC has no health effects; it is an analytic method used to measure the variety of bacteria that are common in water. The lower the concentration of bacteria in drinking water, the better maintained the water system is.	HPC measures a range of bacteria that are naturally present in the environment	n/a
OC	Hexachlorobenzene	0.001	Liver or kidney problems; reproductive difficulties; increased risk of cancer	Discharge from metal refineries and agricultural chemical factories	zero
OC	Hexachlorocyclopentadien[e]	0.05	Kidney or stomach problems	Discharge from chemical factories	0.05
IOC	Lead	TT[g]; action level=0.015	Infants and children: delays in physical or mental development; children could show slight deficits in attention span and learning abilities; adults: kidney problems; high blood pressure	Corrosion of household plumbing systems; erosion of natural deposits	zero
M	Legionella	TT[c]	Legionnaire's Disease, a type of pneumonia	Found naturally in water; multiplies in heating systems	zero
OC	Lindane	0.0002	Liver or kidney problems	Runoff/leaching from insecticide used on cattle, lumber, gardens	0.0002
IOC	Mercury (inorganic)	0.002	Kidney damage	Erosion of natural deposits; discharge from refineries and factories; runoff from landfills and croplands	0.002
OC	Methoxychlor	0.04	Reproductive difficulties	Runoff/leaching from insecticide used on fruits, vegetables, alfalfa, livestock	0.04

deposition. Some VOCs do not degrade quickly and can remain in groundwater for years and even decades. VOCs are of concern not only because they contaminate ground-water but also because their presence in groundwater signals that soil and other conditions favor VOCs reaching the groundwater.

TABLE 4.1

Type	Contaminant	MCL or TT[a] (mg/L)[b]	Potential health effects from exposure above the MCL	Common sources of contaminant in drinking water	Public health goal
IOC	Nitrate (measured as nitrogen)	10	Infants below the age of six months who drink water containing nitrate in excess of the MCL could become seriously ill and, if untreated, may die. Symptoms include shortness of breath and blue-baby syndrome.	Runoff from fertilizer use; leaching from septic tanks, sewage; erosion of natural deposits	10
IOC	Nitrite (measured as nitrogen)	1	Infants below the age of six months who drink water containing nitrite in excess of the MCL could become seriously ill and, if untreated, may die. Symptoms include shortness of breath and blue-baby syndrome.	Runoff from fertilizer use; leaching from septic tanks, sewage; erosion of natural deposits	1
OC	Oxamyl (vydate)	0.2	Slight nervous system effects	Runoff/leaching from insecticide used on apples, potatoes, and tomatoes	0.2
OC	Pentachlorophenol	0.001	Liver or kidney problems; increased cancer risk	Discharge from wood preserving factories	zero
OC	Picloram	0.5	Liver problems	Herbicide runoff	0.5
OC	Polychlorinated biphenyls (PCBs)	0.0005	Skin changes; thymus gland problems; immune deficiencies; reproductive or nervous system difficulties; increased risk of cancer	Runoff from landfills; discharge of waste chemicals	zero
R	Radium 226 and radium 228 (combined)	5 pCi/L	Increased risk of cancer	Erosion of natural deposits	zero
IOC	Selenium	0.05	Hair or fingernail loss; numbness in fingers or toes; circulatory problems	Discharge from petroleum refineries; erosion of natural deposits; discharge from mines	0.05
OC	Simazine	0.004	Problems with blood	Herbicide runoff	0.004
OC	Styrene	0.1	Liver, kidney, or circulatory system problems	Discharge from rubber and plastic factories; leaching from landfills	0.1
OC	Tetrachloroethylene	0.005	Liver problems; increased risk of cancer	Discharge from factories and dry cleaners	zero
IOC	Thallium	0.002	Hair loss; changes in blood; kidney, intestine, or liver problems	Leaching from ore-processing sites; discharge from electronics, glass, and drug factories	0.0005
OC	Toluene	1	Nervous system, kidney, or liver problems	Discharge from petroleum factories	1
M	Total coliforms (including fecal coliform and E. coli)	5.0%[d]	Not a health threat in itself; it is used to indicate whether other potentially harmful bacteria may be present[e]	Coliforms are naturally present in the environment as well as feces; fecal coliforms and E. coli only come from human and animal fecal waste.	zero
DBP	Total trihalomethanes (TTHMs)	0.10 0.080 after 12/31/03	Liver, kidney, or central nervous system problems; increased risk of cancer	By-product of drinking water disinfection	n/a[f]
OC	Toxaphene	0.003	Kidney, liver, or thyroid problems; increased risk of cancer	Runoff/leaching from insecticide used on cotton and cattle	zero
OC	2,4,5-TP (silvex)	0.05	Liver problems	Residue of banned herbicide	0.05
OC	1,2,4-Trichlorobenzene	0.07	Changes in adrenal glands	Discharge from textile finishing factories	0.07
OC	1,1,1-Trichloroethane	0.2	Liver, nervous system, or circulatory problems	Discharge from metal degreasing sites and other factories	0.20
OC	1,1,2-Trichloroethane	0.005	Liver, kidney, or immune system problems	Discharge from industrial chemical factories	0.003
OC	Trichloroethylene	0.005	Liver problems; increased risk of cancer	Discharge from metal degreasing sites and other factories	zero
M	Turbidity	TT[c]	Turbidity is a measure of the cloudiness of water. It is used to indicate water quality and filtration effectiveness (e.g., whether disease-causing organisms are present). Higher turbidity levels are often associated with higher levels of disease-causing micro-organisms such as viruses, parasites, and some bacteria. These organisms can cause symptoms such as nausea, cramps, diarrhea, and associated headaches.	Soil runoff	n/a
R	Uranium	30 ug/L as of 12/08/03	Increased risk of cancer, kidney toxicity	Erosion of natural deposits	zero
OC	Vinyl chloride	0.002	Increased risk of cancer	Leaching from PVC pipes; discharge from plastic factories	zero

According to Zogorski et al., the most frequently detected groups of VOCs in aquifers are the trihalomethanes and organic solvents. Trihalomethanes are VOCs that are used as solvents and in refrigeration. Organic solvents are substances containing carbon that dissolve other substances. They include chloroform and alcohol but do not include water.

TABLE 4.1

Drinking water contaminants, their sources, and potential health effects, 2003 [CONTINUED]

Type	Contaminant	MCL or TT[a] (mg/L)[b]	Potential health effects from exposure above the MCL	Common sources of contaminant in drinking water	Public health goal
M	Viruses (enteric)	TT[c]	Gastrointestinal illness (e.g., diarrhea, vomiting, cramps)	Human and animal fecal waste	zero
OC	Xylenes (total)	10	Nervous system damage	Discharge from petroleum factories; discharge from chemical factories	10

Type Legend:
D=Disinfectant
DBP=Disinfection by product
OC=Inorganic chemical
M=Microorganism
OC=Organic chemical
R=Radionuclides
Notes:
[a]Definitions
- Maximum Contaminant Level Goal (MCLG)—The level of a contaminant in drinking water below which there is no known or expected risk to health. MCLGs allow for a margin of safety and are non-enforceable public health goals.
- Maximum Contaminant Level (MCL)—The highest level of a contaminant that is allowed in drinking water. MCLs are set as close to MCLGs as feasible using the best available treatment technology and taking cost into consideration. MCLs are enforceable standards.
- Maximum Residual Disinfectant Level Goal (MRDLG)—The level of a drinking water disinfectant below which there is no known or expected risk to health. MRDLGs do not reflect the benefits of the use of disinfectants to control microbial contaminants.
- Maximum Residual Disinfectant Level (MRDL)—The highest level of a disinfectant allowed in drinking water. There is convincing evidence that addition of a disinfectant is necessary for control of microbial contaminants.
- Treatment Technique (TT)—A required process intended to reduce the level of a contaminant in drinking water.
[b]Units are in milligrams per liter (mg/L) unless otherwise noted. Milligrams per liter are equivalent to parts per million (ppm).
[c]EPA's surface water treatment rules require systems using surface water or ground water under the direct influence of surface water to (1) disinfect their water, and (2) filter their water or meet criteria for avoiding filtration so that the following contaminants are controlled at the following levels:
- Cryptosporidium (as of 1/1/02 for systems serving >10,000 and 1/14/05 for systems serving <10,000) 99% removal.
- Giardia lamblia: 99.9% removal/inactivation.
- Viruses: 99.99% removal/inactivation.
- Legionella: No limit, but EPA believes that if Giardia and viruses are removed/inactivated, Legionella will also be controlled.
- Turbidity: At no time can turbidity (cloudiness of water) go above 5 nephelolometric turbidity units (NTU); systems that filter must ensure that the turbidity go no higher than 1 NTU (0.5 NTU for conventional or direct filtration) in at least 95% of the daily samples in any month. As of January 1, 2002, for systems servicing >10,000, and January 14, 2005, for systems servicing <10,000, turbidity may never exceed 1 NTU, and must not exceed 0.3 NTU in 95% of daily samples in any month.
- HPC: No more than 500 bacterial colonies per milliliter.
- Long Term 1 Enhanced Surface Water Treatment (Effective Date: January 14, 2005); Surface water systems or (GWUDI) systems serving fewer than 10,000 people must comply with the applicable Long Term 1 Enhanced Surface Water Treatment Rule provisions (e.g. turbidity standards, individual filter monitoring, Cryptosporidium removal requirements, updated watershed control requirements for unfiltered systems).
- Filter Backwash Recycling: The Filter Backwash Recycling Rule requires systems that recycle to return specific recycle flows through all processes of the system's existing conventional or direct filtration system or at an alternate location approved by the state.
[d]No more than 5.0% samples total coliform-positive in a month. (For water systems that collect fewer than 40 routine samples per month, no more than one sample can be total coliform-positive per month.) Every sample that has total coliform must be analyzed for either fecal coliforms or E. coli if two consecutive TC-positive samples, and one is also positive for E. coli fecal coliforms, system has an acute MCL violation.
[e]Fecal coliform and E. coli are bacteria whose presence indicates that the water may be contaminated with human or animal wastes. Disease-causing microbes (pathogens) in these wastes can cause diarrhea, cramps, nausea, headaches, or other symptoms. These pathogens may pose a special health risk for infants, young children, and people with severely compromised immune systems.
[f]Although there is no collective MCLG for this contaminant group, there are individual MCLGs for some of the individual contaminants
- Haloacetic acids: dichloroacetic acid (zero); trichloroacetic acid (0.3 mg/L)
- Trihalomethanes: bromodichloromethane (zero); bromoform (zero); dibromochloromethane (0.06 mg/L)
[g]Lead and copper are regulated by a Treatment Technique that requires systems to control the corrosiveness of their water. If more than 10% of tap water samples exceed the action level, water systems must take additional steps. For copper, the action level is 1.3 mg/L, and for lead is 0.015 mg/L.
[h]Each water system must certify, in writing, to the state (using third-party or manufacturers certification) that when it uses acrylamide and/or epichlorohydrin to treat water, the combination (or product) of dose and monomer level does not exceed the levels specified, as follows: Acrylamide=0.05% dosed at 1 mg/L (or equivalent); Epichlorohydrin=0.01% dosed at 20 mg/L (or equivalent).

SOURCE: "EPA National Primary Drinking Water Standards," U.S. Environmental Protection Agency, Office of Water, June 2003, http://www.epa.gov/safewater/consumer/pdf/mcl.pdf (accessed January 5, 2007)

Zogorski et al. note that VOCs are detected frequently in domestic and public wells, but that only 1% to 2% of the samples taken from these wells had VOC concentrations of potential human-health concern.

Sources of Groundwater Contaminants

In 2000 the EPA requested that states identify the major sources that potentially threaten groundwater in each state. Figure 4.9 shows the results of that survey. Thirty-nine states rated USTs as the most serious threat to their groundwater quality. Septic systems, landfills, industrial facilities, agriculture, and pesticides were also important contamination sources.

LEAKING UNDERGROUND STORAGE TANKS. Leaking underground storage tanks (LUSTs) have been identified by the EPA as the leading source of groundwater contamination since the mid-1990s and were cited as such in the *2000 National Water Quality Inventory* and as a source of VOC groundwater contamination by Zogorski et al.

In general, most underground storage tanks (USTs) are found at commercial and industrial facilities in the more heavily developed urban and suburban areas. USTs are used to store gasoline, hazardous and toxic chemicals, and diluted wastes. Gasoline leaking from UST systems at service stations is one of the most common causes of

FIGURE 4.9

Major sources of groundwater contamination, 2000

[Number reporting on top ten contaminant sources]

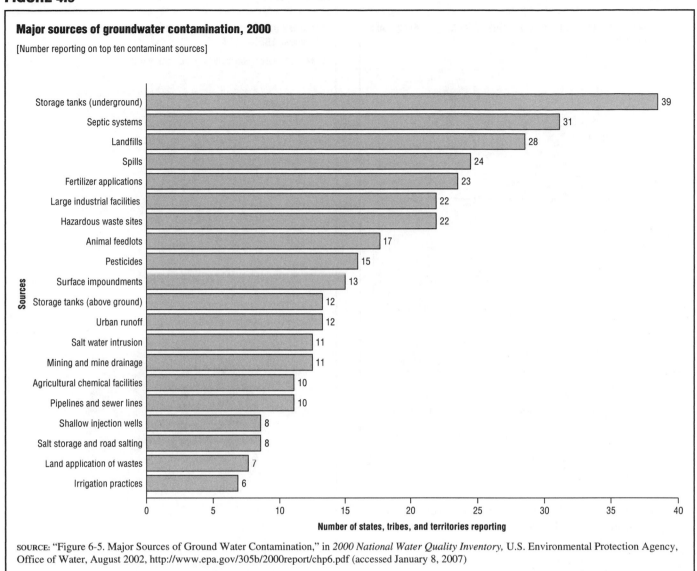

SOURCE: "Figure 6-5. Major Sources of Ground Water Contamination," in *2000 National Water Quality Inventory,* U.S. Environmental Protection Agency, Office of Water, August 2002, http://www.epa.gov/305b/2000report/chp6.pdf (accessed January 8, 2007)

groundwater contamination. The primary causes of leakage in USTs are faulty installation and corrosion of tanks and pipelines.

At one time, USTs were made of steel, which eventually rusted and disintegrated, releasing their contents into the soil. This led to the discovery that a contaminant in the ground is likely to become a contaminant of groundwater. The Sierra Club reports in the news release "Leaking Underground Storage Tanks Continue to Contaminate Groundwater" (April 19 , 2005, http://www.sierraclub.org/pressroom/releases/pr2005-04-19.asp) that one gallon of gasoline can contaminate one million gallons of water. The fuel additive methyl tertiary butyl ether (MTBE), which is a VOC, is particularly troublesome because it migrates quickly through soils into groundwater, and small amounts can render groundwater undrinkable. Figure 4.10 shows how groundwater can be contaminated by LUSTs.

In 1986 Subtitle I of the Solid Waste Disposal Act created the Underground Storage Tank (UST) Program under the management of the EPA. In 1988 the EPA issued "comprehensive and stringent" rules that required devices to detect leaks, modification of tanks to prevent corrosion, regular monitoring, and immediate cleanup of leaks and spills. By December 1998 existing tanks had to be upgraded to meet those standards, replaced with new tanks, or closed. Existing tanks were to be replaced with expensive tanks made of durable, noncorrosive materials.

In testimony before the Subcommittee on Environment and Hazardous Materials, Committee on Energy and Commerce, U.S. House of Representatives, John Stephenson (March 5, 2003, http://www.gao.gov/new.items/d03529t.pdf), the director of the Natural Resources and Environment, stated that as of December 2002 at least 19% to 26% of states still had problems with LUSTs. Although 89% of the 693,107 tanks subject to

FIGURE 4.10

Groundwater contamination as a result of leaking underground storage tanks

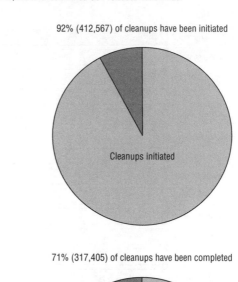

SOURCE: "Figure 6-3. Ground Water Contamination as a Result of Leaking Underground Storage Tanks," in *2000 National Water Quality Inventory*, U.S. Environmental Protection Agency, Office of Water, August 2002, http://www.epa.gov/305b/2000report/chp6.pdf (accessed January 5, 2007)

FIGURE 4.11

Progress on cleaning up underground storage tank (UST) releases, 1988–2004

Of 447,233 total confirmed UST releases nationwide:

92% (412,567) of cleanups have been initiated

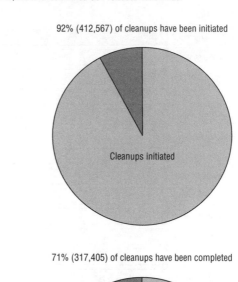

71% (317,405) of cleanups have been completed

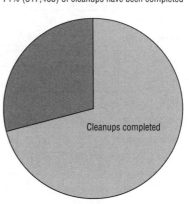

29% (129,827) of cleanups have yet to be completed and remain in the national backlog

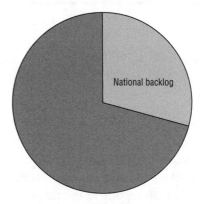

SOURCE: "Progress On Cleaning Up Underground Storage Tank (UST) Releases," in *Cleaning Up Leaks from Underground Storage Tanks*, U.S. Environmental Protection Agency, Office of Underground Storage Tanks, February 2005, http://www.epa.gov/OUST/pubs/cleanup_brochure.pdf (accessed January 8, 2007)

UST rules had leak prevention and detection equipment installed, more than 200,000 tanks were not being operated or maintained properly. The states reported that because of inadequate operation and maintenance of the leak detection equipment, even those tanks with the new equipment continued to leak. To address the problems, Stephenson recommended that Congress provide states more funds from the UST federal trust fund that was created in 1986 to ensure improved training, inspections, and enforcement efforts. In addition, Stephenson suggested that Congress require the states to inspect tanks at least every three years and provide the EPA and the states with additional enforcement authorities.

UST owners and operators must also meet financial responsibility requirements that ensure they will have the resources to pay for costs associated with cleaning up releases and compensating third parties. Many states provide financial assurance funds that help UST owners meet the financial requirements.

As of the end of 2004, 447,233 releases of contaminants from corroded USTs had been confirmed. (See Figure 4.11.) The EPA estimates in the fact sheet *Cleaning Up Leaks from Underground Storage Tanks* (February 2005, http://www.epa.gov/OUST/pubs/cleanup_brochure.pdf) that about half

of these releases reached groundwater. According to the EPA, 92% (412,567) of cleanups had been initiated, 71%

(317,405) had been completed, and 29% (129,827) remained in the national backlog of cleanups in 2004.

In *Environmental Protection: More Complete Data on Continued Emphasis on Leak Prevention Could Improve EPA's Underground Storage Tank Program* (November 2005, http://www.gao.gov/new.items/d0645.pdf), the U.S. Government Accountability Office (GAO) notes that as of early 2005 cleanup had not yet begun on over thirty-two thousand LUSTs. The GAO suggests that the EPA's UST Program could benefit from more specific data on abandoned tanks and that the EPA obtain data on USTs that the states compile.

On August 8, 2005, President George W. Bush signed the Energy Policy Act of 2005, which contained amendments to Subtitle I of the Solid Waste Disposal Act, the original legislation that created the UST Program. These amendments were titled the Underground Storage Tank Compliance Act of 2005, required major changes to the UST Program, and were aimed at preventing releases from USTs. The new legislation also expanded eligible uses of the LUST Trust Fund and included provisions regarding inspections, containment, financial responsibility, and cleanup.

LANDFILLS AND SURFACE IMPOUNDMENTS. In 2000 septic systems and landfills were the second and third largest sources of groundwater contamination, respectively. (See Figure 4.9.) Landfills are areas set aside for disposal of garbage, trash, and other municipal wastes. Early environmental regulation aimed at reducing air and surface water pollution called for disposing of solid wastes—including industrial wastes—underground and gave little consideration to the potential for groundwater contamination. Landfills were generally situated on land considered to have no other use. Many of the disposal sites were nothing more than large holes in the ground, abandoned gravel pits, old strip mines, marshlands, and sinkholes.

The leachate (the liquid that percolates through the waste materials) from landfills contains contaminants that can easily pollute groundwater when disposal areas are not properly lined. Landfills built and operated before the passage of the 1976 Resource Conservation and Recovery Act (RCRA; also known as the Solid Waste Disposal Act) are believed to represent the greatest risk. The RCRA was enacted to protect human health and the environment by establishing a regulatory framework to investigate and address past, present, and future environmental contamination of groundwater and other media. The adoption of these new standards in 1976 forced many landfills to close, as they could not meet the RCRA's safety standards, but in many cases the garbage dumped in them while in operation remains in place and is a threat to groundwater.

Surface impoundments are the industrial equivalent of landfills for liquids and usually consist of human-made

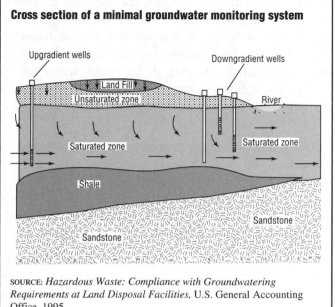

FIGURE 4.12

Cross section of a minimal groundwater monitoring system

SOURCE: *Hazardous Waste: Compliance with Groundwatering Requirements at Land Disposal Facilities,* U.S. General Accounting Office, 1995

pits, lagoons, and ponds that receive treated or untreated wastes directly from the discharge point. They may also be used to store chemicals for later use, to wash or treat ores, or to treat water for further use. Most are small, less than one acre, but some industrial and mining impoundments may be as large as a thousand acres.

Before the RCRA most impoundments were not lined with a synthetic or impermeable natural material, such as clay, to prevent liquids from leaching into the ground. This is particularly important because impoundments are often located over aquifers that are used as sources of drinking water and that may discharge into nearby surface water. Aquifers located under nonlined impoundments are vulnerable to contamination.

Since the passage of the RCRA, landfills and surface impoundments have been required to adhere to increasingly stringent regulations for site selection, construction, operation, and groundwater monitoring to avoid contaminating groundwater. Prevention of groundwater contamination is largely the responsibility of state and local government. Examples of the more stringent requirements are landfill liners and groundwater monitoring.

Figure 4.12 shows the positioning of a lined landfill in the unsaturated zone. Groundwater monitoring is accomplished by sampling the water in an upgradient well to assess the quality of the groundwater before it passes the landfill site. Downgradient wells are used to assess the quality of groundwater at various levels after it flows past the landfill site.

HAZARDOUS WASTE SITES. In general, hazardous wastes are substances with the potential to harm human health and the environment. Hazardous waste is an unavoidable byproduct of an industrial society, because many chemicals are used to manufacture goods. Hazardous waste generators can be large industries, such as automobile manufacturers, or small neighborhood businesses, such as the local photo shop. Although the quantity of hazardous waste can be reduced through innovation and good management, it is impossible to eliminate all hazardous residue because of the demand for goods.

According to the National Solid Wastes Management Association (November 8, 2006, http://wastec.isproductions .net/webmodules/webarticles/anmviewer.asp?a=1121), 30.2 million tons of RCRA hazardous waste were generated in 2003. Texas generated the most at 22% of the total hazardous waste, followed by Louisiana (15%) and Kentucky (8%).

Contamination of groundwater with hazardous waste is frequently the result of historic indiscriminate waste disposal in landfills, impoundments, and dumps. Sites that handle hazardous waste or a mix of hazardous and nonhazardous waste are subject to strict controls.

When a waste site is found to be so badly contaminated with hazardous waste that it represents a serious threat to human health (e.g., contamination of groundwater used for drinking with known carcinogens—cancer-causing agents), it is placed on the National Priorities List, which was established by the Comprehensive Environmental Response, Compensation, and Liability Act of 1980 (CERCLA; commonly known as the Superfund). Sites placed on the Superfund list are eligible for federal intervention and cleanup assistance. The EPA reports in "NPL Site Totals by Status and Milestone" (http://www .epa.gov/superfund/ sites/query/queryhtm/npltotal.htm) that as of March 2007, 1,243 sites were listed on the National Priorities List. Most were general sites such as industrial and municipal landfills and military bases.

INJECTION WELLS. An injection well is any bored, drilled, driven shaft, or dug hole that is deeper than it is wide that is used for the disposal of waste underground. In "Classes of Injection Wells" (February 28, 2006, http://www.epa.gov/safewater/uic/classes.html), the EPA identifies five classes of injection wells of the Underground Injection Control (UIC) Program:

- Class I wells are used to inject hazardous and nonhazardous waste beneath the lowest formation containing an underground source of drinking water (USDW) within a quarter mile of the well bore.

- Class II wells are used to inject fluids associated with oil and natural gas recovery and storage of liquid hydrocarbons.

- Class III wells are used in connection with the solution mining of minerals that are not conventionally mined.

- Class IV wells are used to inject hazardous or radioactive waste into or above a formation that is within a quarter mile of a USDW.

- Class V wells are injection wells not included in Classes I through IV.

Each well class can contaminate groundwater. Classes I through IV have specific regulations and are closely monitored. Class V wells are typically shallow wells used to place a variety of fluids underground.

The EPA reports in "Shallow Injection Wells (Class V)" (November 24, 2006, http://www.epa.gov/safewater/ uic/classv/index.html) that in 2006 there were more than five hundred thousand Class V injection wells in the United States. These wells are found in every state, especially in unsewered areas. There are many types of Class V wells, including large-capacity cesspools, motor vehicle waste disposal systems, storm water drainage wells, large-capacity septic systems, aquifer remediation wells, and many others. The waste entering these wells is not treated. Certain types of these wells have great potential to have high concentrations of contaminants that might endanger groundwater.

Class V injection wells are regulated by the UIC Program under the authority of the Safe Drinking Water Act (SDWA). Class V wells are "authorized by rule," which means that they do not require a permit if they comply with UIC Program requirements and do not endanger underground sources of drinking water. In December 1999 the EPA adopted regulations addressing Class V wells that were large-capacity cesspools and motor vehicle waste disposal wells. Under these regulations:

- New cesspools were prohibited as of April 2000.

- Existing cesspools had to be phased out by April 2005.

- New motor vehicle waste disposal wells were prohibited.

- Existing wells in regulated areas were to be phased out in groundwater protection areas identified in state source water assessment programs.

AGRICULTURE. As in surface water contamination, agricultural practices play a major role in groundwater contamination. Agricultural practices that have the potential to contaminate groundwater include fertilizer and pesticide applications, animal feedlots, irrigation practices, agricultural chemical facilities, and drainage wells. Contamination can result from routine applications, spillage or misuse of pesticides and fertilizers during handling and storage, manure storage and spreading, improper storage of chemicals, irrigation practices, and irrigation return drains serving as direct conduits to groundwater. Fields with overapplied or misapplied fertilizer and pesticides can introduce nitrogen, pesticides, and other contaminants into groundwater. Animal feedlots often have impoundments from which wastes (bacteria, nitrates, and total and dissolved solids) may infiltrate groundwater.

Human-induced salinity in groundwater also occurs in agricultural regions where irrigation is used extensively. Irrigation water continually flushes nitrate-related compounds from fertilizers into the shallow aquifers along with high levels of chloride, sodium, and several types of metals. This increases the salinity (dissolved solids) of the underlying aquifers. Overpumping can diminish the water in aquifers to the point where saltwater from nearby coastal areas will intrude into the aquifer. Salinas Valley, California, is an example of the occurrence of saltwater intrusion. Eleven states identified saltwater intrusion as a major source of groundwater contamination in their 2000 305(b) reports to the EPA. (See Figure 4.9.)

SEPTIC SYSTEMS. Septic systems were cited as the second most common source of groundwater contamination by thirty-one reporting states. (See Figure 4.9.) Septic systems are on-site waste disposal systems that are used where public sewerage is not available. Septic tanks are used to detain domestic wastes to allow the settling and digestion of solids before the distribution of liquid wastes into permeable leach beds for absorption into soil. Wastewater is digested in the leach beds by organisms in the soil and broken down over time.

According to the EPA (September 20, 2006, http://cfpub.epa.gov/owm/septic/faqs.cfm?program_id=70#359), the U.S. Census Bureau reports that approximately twenty-six million Americans use individual sewage disposal systems. The EPA notes that the use of septic systems varies across the country from about 55% of the population in Vermont to around 10% in California. More than sixty million people nationwide live in homes with septic systems. Improperly constructed and poorly maintained septic systems may cause substantial and widespread contamination to groundwater of nitrogen and disease-causing microbes.

GROUNDWATER CLEANUP

Cleaning up the nation's groundwater is expensive. The costs associated with alternative water supplies, water treatment, and contaminant source removal or remediation are in the millions per site. However, the GAO (May 14, 2004, http://www.gao.gov/new.items/d04787r.pdf) reports that "the net Superfund program appropriations...decreased from $1,757 million to $1,242 million, in constant 2003 dollars, from fiscal year 1993 to fiscal year 2004."

In allocating limited resources, cleanup decisions are based on a cost-benefit analysis that considers such factors as the extent of the problem, the potential health effects, and the alternatives, if any. If the pollution is localized, it may be more practical to simply shut down the contaminated wells and find water elsewhere. Cleanup options range from capping a section of an aquifer with a layer of impermeable clay to prevent more pollution, to more complex (and expensive) methods, such as pumping out and treating the water and then returning it to the aquifer.

GROUNDWATER PROTECTION
States' Role

Prevention of groundwater contamination is largely the responsibility of state and local government. In 1991 the EPA established a national groundwater protection strategy to place greater emphasis on comprehensive state management of groundwater resources. The EPA recognized that the wide range of land-use practices that can adversely affect groundwater quality is most effectively managed at the state and local levels. The states use three basic approaches to protect groundwater and address the problems of contaminants and contamination sources:

- Nondegradation policies that are designed to protect groundwater quality at its existing level.

- Limited degradation policies that involve setting up water quality standards to protect groundwater. These standards set maximum contamination levels for chemicals and bacteria and establish guidelines for taste, odor, and color of the water.

- Groundwater classification systems that are similar to the classification systems for surface waters established under the Clean Water Act and its amendments.

These classification systems are used by state officials to determine which aquifers should receive higher or lower priorities for protection and cleanup. High-priority areas include recharge areas, which affect large quantities of water, or public water supplies, where pollution affects drinking water.

The most important benefit derived from comprehensive groundwater management approaches is the ability to establish coordinated priorities among the many groups involved in groundwater management. The following key components are common to successful state programs:

- Enacting legislation

- Publicly announcing protection regulations

- Establishing interagency coordination with surface water and other programs

- Performing groundwater mapping and classification

- Monitoring groundwater quality

- Developing comprehensive data management systems

- Adopting and implementing prevention and remediation programs

Federal Role

Federal laws, regulations, and programs since the 1970s have reflected the growing recognition of the need to protect the nation's groundwater and use it wisely. Table 4.2 summarizes major federal legislation affecting groundwater. The Federal Water Pollution Control Act (Clean Water Act) in 1972 and the SDWA in 1974 began the federal role in groundwater protection. The passage of the RCRA in 1976 and CERCLA (or Superfund) in 1980 cemented the federal government's current focus on groundwater remediation. Since the passage of these acts, the federal government has directed billions of dollars in private and public money and resources toward the cleanup of contaminated groundwater at Superfund, RCRA corrective action facilities, and LUSTs.

TABLE 4.2

Federal laws administered by the EPA affecting groundwater

Clean Water Act (CWA)

Groundwater protection is addressed in Section 102 of the CWA, providing for the development of federal, state, and local comprehensive programs for reducing, eliminating, and preventing groundwater contamination.

Safe Drinking Water Act (SDWA)

Under the SDWA, EPA is authorized to ensure that water is safe for human consumption. To support this effort, SDWA gives EPA the authority to promulgate Maximum Contaminant Levels (MCLs) that define safe levels for some contaminants in public drinking water supplies. One of the most fundamental ways to ensure consistently safe drinking water is to protect the source of that water (i.e., groundwater). Source water protection is achieved through four programs: the Wellhead Protection Program (WHP), the Sole Source Aquifer Program, the Underground Injection Control (UIC) Program, and, under the 1996 Amendments, the Source Water Assessment Program.

Resource Conservation and Recovery Act (RCRA)

The intent of RCRA is to protect human health and the environment by establishing a comprehensive regulatory framework for investigating and addressing past, present, and future environmental contamination or groundwater and other environmental media. In addition, management of underground storage tanks is also addressed under RCRA.

Comprehensive Environmental, Response, Compensation, and Liability Act (CERCLA)

CERCLA provides a federal "superfund" to clean-up soil and groundwater contaminated by uncontrolled or abandoned hazardous waste sites as well as accidents, spill, and other emergency releases of pollutants and contaminants into the environment. Through the Act, EPA was given power to seek out those parties responsible for any release and assure their cooperation in the clean-up. The program is designed to recover costs, when possible, from financially viable individuals and companies when the clean-up is complete.

Federal Insecticide, Fungicide, and Rodenticide Act (FIFRA)

FIFRA protects human health and the environment from the risks of pesticide use by requiring the testing and registration of all chemicals used as active ingredients of pesticides and pesticide products. Under the Pesticide Management Program, states and tribes wishing to continue use of chemicals of concern are required to prepare a prevention plan that targets specific areas vulnerable to groundwater contamination.

Superfund Amendments and Reauthorization Act (SARA)

SARA reauthorized CERCLA in 1986 to continue cleanup activities around the country Several site-specific amendments, definitions clarifications, and technical requirements were added to the legislation, including additional enforcement authorities.

SOURCE: Adapted from "Federal Laws Administered by EPA Affecting Ground Water," in *Safe Drinking Water Act, Section 1429 Ground Water Report to Congress*, U.S. Environmental Protection Agency, Office of Water, October 1999, http://www.gwpc.org/e-Library/Documents/GW_Report_to_Congress.pdf (accessed January 8, 2007)

CHAPTER 5
DRINKING WATER—SAFETY ON TAP

HOW MUCH WATER DO AMERICANS USE?

In the American Water Works Association's benchmark study, *Residential End Uses of Water Study* (1999, http://www.awwarf.org/research/topicsandprojects/exec Sum/241.aspx), Peter W. Mayer et al. report the results of their study on residential end uses of water in twelve hundred single-family homes in twelve North American locations from 1996 to 1998. Mayer et al. reveal that, on average, Americans on community water supplies use about one hundred gallons of water per person per day. People with private wells use slightly less. About sixty-nine gallons per day are used indoors and the rest is used outdoors. According to the University of North Carolina at Chapel Hill, in "Beverage Intake in the United States" (May 5, 2006, http://www.cpc.unc.edu/projects/beverage/beverage_home.html), of this daily supply, only a small portion—slightly more than one-third of a gallon—is consumed as drinking water.

Residential water consumers use most water for purposes other than drinking, such as toilet flushing, bathing, cooking, and cleaning. In the United States significant amounts of water are used for kitchen and laundry appliances, such as garbage disposals, clothes washers, and automatic dishwashers; for automobile washing; and for lawn and garden watering. Additional community use includes firefighting, fountains, public swimming pools, and watering of public parks and landscaping.

DRINKING WATER SOURCES

The two primary sources of drinking water are surface freshwater and groundwater. In 2000 more than half (63%; 27.3 billion gallons per day [Bgal/d] of 43.3 Bgal/d) of public-supply water withdrawals were from surface water sources (e.g., lakes, rivers, and reservoirs). (See Table 2.3 in Chapter 2.) The remaining water withdrawals (37%; 16 Bgal/d) were supplied with water that came from groundwater stored in aquifers. Aquifers are underground geologic formations that consist of layers of sand and porous rock that are saturated with water. Aquifer water is obtained from wells and springs. The only other source of drinking water is desalinated seawater, which is used in only a few locations around the world and provides little of the total amount of drinking water worldwide.

PUBLIC AND PRIVATE WATER SUPPLIES
Public-Supply Water

As described in Chapter 2, public-supply water use is water withdrawn by public and private water suppliers (utility companies) and delivered for public uses: domestic, commercial, industrial, and thermoelectric power uses. It may be used for public services such as filling public pools, watering vegetation in parks, supplying public buildings, firefighting, and street washing. The latest data available from the U.S. Geological Survey show that 43.3 Bgal/d were supplied to users in 2000 by water utility companies (public supply). (See Table 2.2 in Chapter 2.) The rest of the water was self-supplied—that is, the water was withdrawn from groundwater or surface water sources by the users, not by water utility companies.

Public-supply water systems (which can be publicly or privately owned) have at least fifteen service connections or serve at least twenty-five people per day for sixty days of the year. According to the U.S. Environmental Protection Agency (EPA) report *Factoids: Drinking Water and Ground Water Statistics for 2005* (December 2006, http://www.epa.gov/ogwdw/data/pdfs/statistics_data _factoids_2005.pdf), there were 156,582 of these systems of varying size in the United States in 2005. (See Table 5.1.) The amount and type of treatment provided varies with source and quality. For example, some public systems using a groundwater source require no treatment, whereas others may need to disinfect the water or apply additional treatment.

TABLE 5.1

Types of public water systems, by water source and population served, 2005

Water source	Systems	Population served	Percent of systems	Percent of population
Community water systems				
Groundwater	40,018	89,539,197	77%	32%
Surface water	11,737	191,130,147	23%	68%
Total	51,755	280,669,344	100%	100%
Nontransient noncommunity water systems				
Groundwater	18,438	5,410,376	97%	90%
Surface water	607	611,002	3%	10%
Total	19,045	6,021,378	100%	100%
Transient noncommunity water systems				
Groundwater	83,930	11,305,555	98%	93%
Surface water	1,852	801,399	2%	7%
Total	85,782	12,106,954	100%	100%
Total				
Groundwater	142,386	106,255,128		
Surface Water	14,196	192,542,548		
Total	156,582	298,797,676		

SOURCE: Adapted from "Water Source," in *Factoids: Drinking Water and Ground Water Statistics for 2005*, U.S. Environmental Protection Agency, Office of Water, December 2006, http://www.epa.gov/safewater/data/pdfs/statistics_data_factoids_2005.pdf (accessed January 8, 2007)

There are three types of public water systems. Figure 5.1 shows a flowchart of drinking water systems, including public water systems. Table 5.1 displays their similarities and differences. Community water systems are those that supply water to the same population year-round. Most people in the United States are served by community water systems. In 2005 there were 51,755 community water systems serving 280.7 million (94%) out of 298.8 million people in the United States. Of these systems, 40,018 (77%) accessed groundwater for their water supply. Nevertheless, they served only 89.5 million (32%) of the community water system population. In 2005, 191.1 million (68%) of that population was served by 11,737 community water systems (23%) that access surface water for their water supply.

Nontransient noncommunity water systems are the second type of public water system. They serve the public but not the same people year-round. Examples of nontransient noncommunity systems are schools, factories, office buildings, hospitals, and other public accommodations. In 2005 there were 19,045 nontransient noncommunity water systems in the United States serving 6 million people, or 2% of the total U.S. population. (See Table 5.1.) Of these systems, 18,438 (97%) accessed groundwater for their water supply.

Transient noncommunity water systems are the third type of public water system. These are systems that provide water in places such as gas stations or campgrounds, where people do not remain for long periods of time. In 2005 there were 85,782 transient noncommunity water systems serving 12.1 million people, or 4% of the total U.S. population. (See Table 5.1.) Ninety-eight percent (83,930) of transient noncommunity water systems accessed groundwater for their water supply.

The EPA and state health and environment departments regulate public water supplies. Public suppliers are required to ensure that the water meets certain government-defined health standards under the Safe Drinking Water Act (SDWA). This law mandates that all public suppliers test their water regularly to check for the existence of contaminants and treat their water supplies, if necessary, to take out or reduce certain pollutants to levels that will not harm human health.

The EPA notes in *Factoids* that more water systems have groundwater than surface water as a source, but more people drink from a surface water system.

Table 5.2 shows that 149,182 (or 94%) of public water systems were small or very small in 2005, each serving fewer than 3,300 people. (See Table 5.2.) The remaining systems (9,039, or 6%) were comparatively few in number but serviced many more people. The medium-sized systems each provide water to between 3,301 and 10,000 people. The large and very large systems provide water to more than ten thousand people each. Together, the medium, large, and very large public water services provided water for the vast majority of people who drank water from a public supply in the United States in 2005: 261.8 million people.

FIGURE 5.1

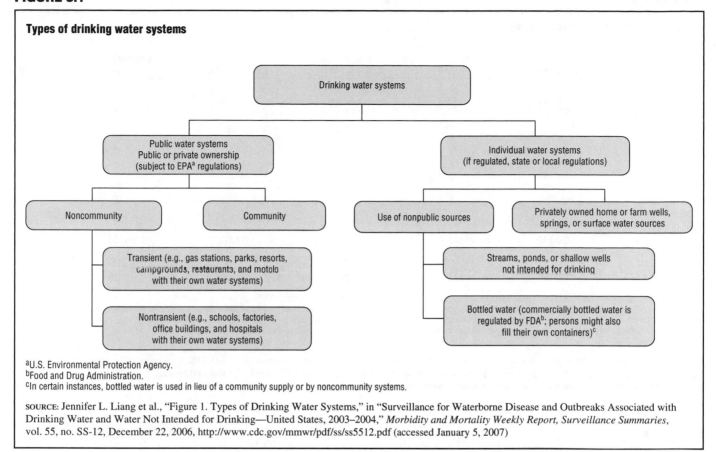

Types of drinking water systems

[a]U.S. Environmental Protection Agency.
[b]Food and Drug Administration.
[c]In certain instances, bottled water is used in lieu of a community supply or by noncommunity systems.

SOURCE: Jennifer L. Liang et al., "Figure 1. Types of Drinking Water Systems," in "Surveillance for Waterborne Disease and Outbreaks Associated with Drinking Water and Water Not Intended for Drinking—United States, 2003–2004," *Morbidity and Mortality Weekly Report, Surveillance Summaries*, vol. 55, no. SS-12, December 22, 2006, http://www.cdc.gov/mmwr/pdf/ss/ss5512.pdf (accessed January 5, 2007)

Private Water Systems

According to the EPA, in *Drinking Water from Household Wells* (January 2002, http://www.epa.gov/safewater/privatewells/pdfs/household_wells.pdf), 15% of Americans obtain their water from private wells, cisterns, and springs. System owners are solely responsible for the quality of the water provided from these sources.

Personal private water supplies, usually wells, are not regulated under the SDWA. Many states, however, have programs designed to help well owners protect their own water supplies. Usually, these state-run programs are not regulatory, but provide safety information. In addition, the EPA is a source of information. This type of information is vital because private wells are often shallower than those used by public suppliers. The more shallow the well, the greater the potential for contamination.

CONTAMINANTS IN DRINKING WATER

Water can dissolve many substances. Pure water rarely occurs in nature, because both surface and groundwater dissolve minerals and other substances in the soil and deposited from the atmosphere. At low levels dissolved contaminants generally are not harmful in drinking water. Removing all contaminants would be extremely expensive and might not provide greater protection of health. The concentration of harmful substances in water is the main determinant in whether the water is safe to drink.

Contaminants in drinking water are grouped into two broad categories: chemical and microbial. Both chemical and microbial contaminants may be naturally occurring or may be caused by human activity. Chemical contaminants include metals, pesticides, synthetic chemical compounds, suspended solids, and other substances. Microbial contaminants include bacteria, viruses, and microscopic parasites. A rather thorough list of drinking water contaminants, their sources, and their potential health effects is shown in Table 4.1 in Chapter 4.

The health effects of drinking contaminated water can occur either over a short or long period. Short-term, or acute, reactions are those that occur within a few hours or days after drinking contaminated water. Acute reactions may be caused by a chemical or microbial contaminant. Long-term, or chronic, effects occur after water with relatively low doses of a pollutant has been consumed for several years or over a lifetime. Most chronic effects are caused by chemical contaminants.

The ability to detect contaminants improved considerably in the late twentieth century. Scientists can now identify specific chemical pollutants in terms of one part contaminant in one billion parts of water. In some cases

TABLE 5.2

Public water system, by size and population served, 2005

	Very small 500 or less	Small 501–3,300	Medium 3,301–10,000	Large 10,001–100,000	Very large >100,000	Total
CWS						
# systems	29,666	14,389	4,748	3,648	386	52,837
Population served	4,925,748	20,851,292	27,514,714	102,747,558	126,304,807	282,344,119
% of systems	56%	27%	9%	7%	1%	100%
% of population	2%	7%	10%	36%	45%	100%
NTNCWS						
# systems	16,348	2,707	102	17		19,174
Population served	2,282,628	2,710,912	557,742	504,915		6,056,197
% of systems	85%	14%	1%	0%	0%	100%
% of population	38%	45%	9%	8%	0%	100%
TNCWS						
# systems	83,351	2,721	111	23	4	86,210
Population served	7,298,704	2,667,051	598,506	604,213	2,994,000	14,162,474
% of systems	97%	3%	0%	0%	0%	100%
% of population	52%	19%	4%	4%	21%	100%
Total # systems	**129,365**	**19,817**	**4,961**	**3,688**	**390**	**158,221**

CWS=Community water system: A public water system that supplies water to the same population year-round.
NTNCWS=Nontransient noncommunity water system: A public water system that regularly supplies water to at least 25 of the same people at least six months per year, but not year-round. Some examples are schools, factories, office buildings, and hospitals which have their own water systems.
TNCWS=Transient noncommunity water system: A public water system that provides water in a place such as a gas station or campground where people do not remain for long periods of time.

SOURCE: "System Size by Population Served," in *Factoids: Drinking Water and Ground Water Statistics for 2005*, U.S. Environmental Protection Agency, Office of Water, December 2006, http://www.epa.gov/safewater/data/pdfs/statistics_data_factoids_2005.pdf (accessed January 8, 2007)

scientists can measure them in parts per trillion. One part per billion (ppb) is equal to one pound in five hundred thousand tons. Although these measurements appear tiny, such small amounts can be significant in terms of health effects.

Chemical Contaminants

All drinking water contains minerals dissolved from the earth. In small amounts some of these minerals are acceptable because they often enhance the quality of the water (e.g., giving it a pleasant taste). A few minerals in small amounts, such as zinc and selenium, contribute to good health. Other naturally occurring minerals are not desirable because they may cause a bad taste or odor (such as excessive amounts of iron, manganese, or sulfur) or because they may be harmful to health (such as boron).

A wide variety of contaminants may cause serious health risks in water supplies. Not all contaminants are found in all water supplies; furthermore, some water supplies have no undesirable contaminants and some supplies have no contaminants that have health significance. Contaminant presence is frequently the result of human activity and may have long-term consequences. Even though harmful levels of microorganisms generally make their presence known quickly by causing illness with fairly obvious symptoms, the effects of some toxic chemicals may not be apparent for months or even years after exposure. Some chemical pollutants are known carcinogens (cancer-causing agents), whereas others are sus-

pected of causing birth defects, miscarriages, and heart disease. In many cases the effects occur only after long-term exposure.

ARSENIC. Arsenic is a naturally occurring element in rocks and soils and is soluble in water. Arsenic has been recognized as a poison for centuries. Recent research, however, shows that humans need arsenic in their diet as a trace element. However, Paolo Boffetta and Fredrik Nyberg report in "Contribution of Environmental Factors to Cancer Risk" (*British Medical Bulletin*, 2003) that too much arsenic can contribute to skin, bladder, and lung cancers after prolonged exposure. Because of this risk, the current maximum contaminant level (MCL) for arsenic in drinking water is ten ppb.

LEAD AND COPPER. Lead is a toxic metal that can cause serious health problems if ingested. Children are particularly at risk because their developing bodies absorb and retain more lead than adult bodies. Low-level exposures can result in a lowered intelligence quotient (IQ), impaired learning and language skills, loss of hearing, reduced attention spans, and poor school performance. High levels damage the brain and central nervous system, interfering with both learning and physical development. Pregnant women are also at risk. Lead can cause miscarriages, premature births, and impaired fetal development.

Unlike many water contaminants, lead has been extensively studied for its prevalence and effects on human health and for ways to eliminate it from the water supply. Lead is rarely found in either surface water or groundwater sources for drinking. This contaminant usually enters the water supply after it leaves the treatment plant or the well.

The major sources of lead exposure are deteriorated lead paint in older houses and dust and soil that are contaminated by old paint and past emissions of leaded gasoline. Plumbing in older buildings has also contributed to overall blood lead levels. Until about 1930 many buildings in the United States had lead pipes in their interior plumbing and for the service connections that linked buildings to the public water supplies. In addition, lead solder was commonly used to connect pipes. There is little lead piping in use any more within buildings, but some lead service piping still exists in one-hundred-year-old inner-city neighborhoods.

Copper pipes replaced lead in most buildings, but the practice of using lead solder to join the pipes continued. The corrosion of lead solder is believed to be the primary cause of most lead in residential water supplies today. Low pH (acidity), low calcium or magnesium levels in the water, and dissolved oxygen can all contribute to corrosion of lead solder. The common practice of grounding electrical equipment to water pipes also accelerates corrosion.

Most commonly, copper pipes are used to supply water from the street to a building, and to supply water to various parts of the building's interior. Although copper is a dependable material, it can be corroded by acidic water.

In June 1991 the EPA (August 24, 2006, http://www.epa.gov/safewater/lcrmr/index.html) established a regulation called the Lead and Copper Rule (LCR) to help control lead and copper in drinking water. The rule requires community and nontransient noncommunity water systems to monitor drinking water at customer taps. If lead concentrations exceed 15 ppb or copper concentrations exceed 1.3 parts per million (ppm) in more than 10% of the customer taps sampled, then action must be taken to control corrosion and possibly replace lead service lines. If lead levels are exceeded, then the public must be informed about how to protect themselves against lead poisoning.

The EPA (July 18, 2006, http://www.epa.gov/fedrgstr/EPA-WATER/2006/July/Day-18/w6250.pdf) proposed changes to the LCR in 2006 that focused on enhancing monitoring, treatment, customer awareness, and lead service line replacement. The changes will also help ensure that drinking water consumers receive meaningful, timely, and useful information to help them limit their exposure to lead in drinking water.

NITRATES. Nitrates are plant nutrients that enter both surface and groundwater primarily from fertilizer runoff, human sewage, and livestock manure, especially from feedlots. Nitrates in drinking water can be an immediate threat to children under six months of age. In some babies high levels of nitrates react with the red blood cells to reduce the blood's ability to transport oxygen.

The MCL for nitrates is set at ten ppm. When nitrate levels exceed this limit, a water supplier must notify the public and provide additional treatment to reduce levels to meet the standards. The number of community water systems with MCL violations for nitrates declined between 1980 and 1998. The highest number of systems in violation was registered in 1985, with 340 community water systems having levels over the limit.

In *Factoids,* the EPA reports that 943 nitrate violations were reported in 540 community and nontransient noncommunity water systems in fiscal year 2005. The number of people served by suppliers cited for nitrate violations in 2005 was 608,472.

Microbial Contaminants

Microbes (bacteria, viruses, and protozoa) are found in untreated surface water sources used for drinking water. Groundwater does not contain microbes unless they have been introduced through pollution of the aquifer. Unless the treatment system fails or contaminated water is introduced accidentally into the distribution system, treated drinking water is normally free of microorganisms or they are present in extremely low levels. When a water source or system is contaminated with human or animal fecal waste, some of the microorganisms may be pathogens (disease-causing organisms). The resulting illnesses can have symptoms that include headache, nausea, vomiting, diarrhea, abdominal pain, and dehydration. Although usually not life threatening, these illnesses can be debilitating and uncomfortable for victims. Extended illness or death may occur among young or elderly individuals or those who are immunocompromised (having weakened immune systems). Immunocompromised people include human immunodeficiency virus and acquired immunodeficiency syndrome patients, those receiving treatment for certain kinds of cancer, organ-transplant recipients, and people on drugs that suppress their immune system.

Waterborne pathogens have been the cause of serious diseases throughout the world. In the United States in the early 1900s, the diseases cholera and typhoid fever were commonly associated with drinking water from public supplies. The practice of water treatment was begun to address this problem by reducing the number of pathogens present in water supply systems below an infective dose. The infective dose is the number of a particular microorganism required to induce disease and is different

FIGURE 5.2

Activities within a watershed that can increase turbidity

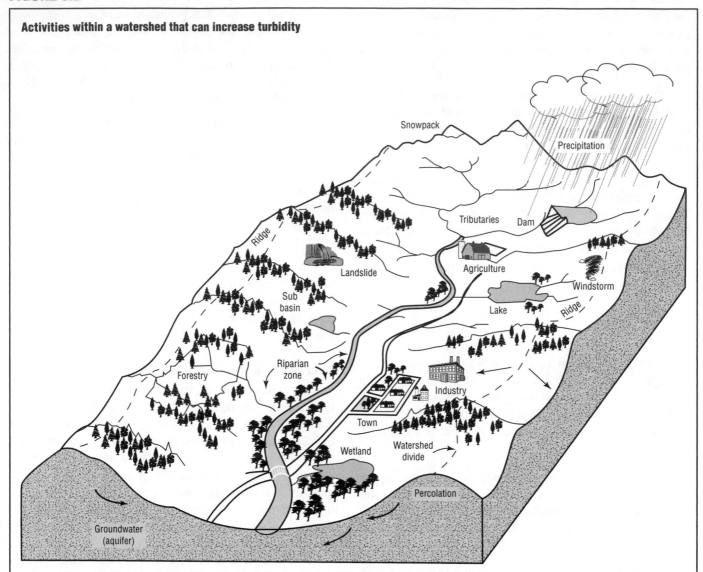

SOURCE: "Figure 2.1. Activities Within a Watershed That Can Increase Turbidity," in *Oregon Watersheds: Many Activities Contribute to Turbidity During Large Storms*, U.S. General Accounting Office, July 1998, http://www.gao.gov/archive/1998/rc98220.pdf (accessed January 4, 2007)

for different microbes. For example, one *Cryptosporidium* protozoan can induce disease, whereas ten thousand to one hundred thousand *Salmonella* bacteria are generally necessary for serious illness to occur.

HOW TURBIDITY AFFECTS MICROBIAL CONTAMINATION. Turbidity is a measure of the clarity of water. Turbidity is caused by suspended matter or impurities that make the water look cloudy. These impurities may include clay, silt, fine organic and inorganic matter, and plankton (minute floating aquatic plants and animals).

Figure 5.2 shows the types of activities in a watershed that cause turbidity in a water source. A watershed is the land area that drains water into a river system or other body of water. Figure 5.2 shows that water falls on the land as rain or snow, and that water runs along the ground and eventually seeps into the groundwater or

surface water. Human activities such as timber harvests, road building, and residential development compact, pave, and clear the soil of much of its vegetation. During storms, rain runs over this land and erodes it, carrying with it impurities.

Turbidity (excessive cloudiness in water) is unappealing and may represent a health concern in drinking water. It interferes with the effectiveness of disinfection, which is the practice of killing pathogens in water by adding certain chemicals (e.g., chlorine or ozone) or exposing the water to ultraviolet light. Microorganisms can find shelter in the particulate matter, reducing their exposure to disinfectants and ultraviolet light. Although turbidity is not a direct indicator of health risk, many studies show a strong relationship between the removal of turbidity and the removal of pathogens.

COLIFORM BACTERIA. Coliform bacteria are a group of closely related, mostly harmless bacteria that live in soil, water, and the intestines of animals. These bacteria are generally divided into two groups: total coliform and fecal coliform. The total coliform group includes all coliform bacteria. The fecal coliform group is a subgroup found in the intestines and fecal waste of warm-blooded animals. There are a few organisms in the fecal coliform group that can be harmful to humans, particularly to children and to immunocompromised people.

The total coliform group is used as a first indicator to assess drinking water quality. This practice began in the early 1900s. It is based on the assumption that because coliform bacteria are always present in sewage from warm-blooded animals (including humans), and pathogens may be present in this same sewage, the presence of coliform bacteria may indicate the potential presence of pathogens. The most common problem caused by fecal pathogens is gastroenteritis, a general illness characterized by diarrhea, nausea, vomiting, and cramps. Even though gastroenteritis is typically not harmful to healthy adults, it can cause serious illness in children and immunocompromised individuals.

Testing the water for each of a wide variety of potential pathogens is difficult and expensive. Testing for total coliform, by comparison, is easy and inexpensive. For this reason, total coliform are used to indicate whether a water system is vulnerable to pathogens. The presence of total coliform in the water distribution system may indicate that the disinfection process is faulty, that a break or leak has occurred in the distribution piping, or that the distribution pipes need to be cleaned. No more than 5% of the drinking water samples collected monthly from a water supplier may be positive for total coliform. All samples that are positive for total coliform are analyzed for the presence of the fecal coliform group or *Escherichia coli* (*E. coli*), a specific member of the fecal coliform group, both of which are more sensitive indicators of sewage pollution.

GIARDIA LAMBLIA **AND** *CRYPTOSPORIDIUM*. *Giardia lamblia* and *Cryptosporidium* are microscopic single-celled protozoa that can infect humans and other warm-blooded animals. They are frequently found in surface waters contaminated with animal or human fecal waste. Both organisms have a life stage called a cyst, in which the organism is dormant and protected by an outer shell that allows it to exist outside a host's body for a long time. If cysts are ingested, they can become active and cause an intestinal illness, the symptoms of which are nausea, vomiting, fever, and severe diarrhea. The symptoms last for several days, and a healthy human can generally rid his or her body of the organisms in one or two months. These two organisms are the most frequent cause of waterborne illness in the United States.

The EPA, in "Drinking Water Contaminants" (November 28, 2006, http://www.epa.gov/safewater/contaminants/index.html), indicates that it requires water suppliers using surface water, or groundwater under the direct influence of surface water, to disinfect their water to control *Giardia* at the 99.9% inactivation and removal level. Groundwater is considered to be under the direct influence of surface water when the geologic formations (usually limestone or fractured bedrock) in which the aquifer lies do not provide adequate natural filtration.

A smaller parasite than *Giardia*, *Cryptosporidium* is fifty times more resistant to chlorine (the most commonly used drinking water disinfectant) than *Giardia* is. Because of its high resistance to chemicals typically used to treat drinking water, it must be physically removed by filtration. The EPA notes in "Drinking Water Contaminants" that as of January 2002 water systems serving ten thousand or more people are required to provide filtration and achieve 99% removal or inactivation of *Cryptosporidium*. This requirement was applied to water systems serving less than ten thousand people in January 2005.

Cryptosporidium was responsible for what many people view as the nation's worst drinking water disaster. In April 1993 residents of Milwaukee were infected with *Cryptosporidium* in the city water supply, which had been turbid for several days. For a week, more than eight hundred thousand residents were without drinkable tap water. By the end of the disaster, fifty people had died and over four hundred thousand people had been infected. Besides the human suffering, the disease outbreak cost millions of dollars in lost wages and productivity.

MODERN WATER TREATMENT

Although the Greek physician Hippocrates is credited with emphasizing the importance of clean water for good health as early as 400 BC (he recommended boiling and straining rainwater), the first recorded observation of the connection between drinking water and the spread of disease came from John Snow, a London physician, in 1849. Snow noted that his patients who were getting their drinking water from one particular well were contracting cholera, whereas patients getting drinking water from other wells were not. His solution to the problem was to remove the handle from the contaminated well's pump so that no one could get water, thereby stopping a cholera epidemic. This event is generally credited as the beginning of modern water treatment.

The most significant water treatment event in the United States was the introduction of chlorine as a disinfectant in water supplies. Adding chlorine to water supplies began in the early 1900s. As towns and cities began implementing this practice, epidemics and incidence of typhoid, cholera, and dysentery were dramatically reduced. From this humble beginning evolved the complex drinking water treatment technology that is currently available.

The multiple-barrier approach is the basis for modern water treatment. This approach recognizes that contaminants reach drinking water through many pathways. Working together, water suppliers and health professionals try to erect as many barriers as possible to prevent contaminants from reaching consumers. These barriers include:

- Protecting the water source from contamination by eliminating or limiting waste discharges to the water source through a variety of protection programs

- Improved contaminant detection methods

- New and ongoing research into contaminants and their effects

- Removing contaminants or reducing contaminant levels through various treatments

- Disinfection

- Elimination of cross connections and breaks in the distribution lines

- Safe plumbing in residences and businesses

The water treatment process begins with choosing the highest quality groundwater or surface water source available and ensuring its continued protection. (For an annotated illustration of the drinking water treatment process, see Figure 5.3.) Groundwater is usually pumped directly into the treatment plant. In many cases, however, because groundwater is naturally filtered as it seeps through layers of rock and soil, disinfection is the only treatment needed before the water is distributed to consumers.

Surface water is transported to the water treatment plant through aqueducts or pipes. A screen at the intake pipe removes debris such as tree branches and trash.

Water suppliers use a variety of treatments to remove contaminants. These treatments are usually arranged in a sequential series of processes called the treatment train. In the plant the water is aerated to eliminate gases and add oxygen. Chemicals may be added to remove undesirable contaminants or to improve the taste. If the water is hard, lime or soda is added to remove the calcium and magnesium. Hard water can clog pipes, stain fixtures, and interfere with soap lathering.

Coagulation or flocculation is typically the next step. Alum, iron salts, or synthetic polymers are added to the water to combine smaller particles into larger particles (floc) to remove contaminants. In the sedimentation basins, the floc settles to the bottom and is removed. Additional treatment may be required if the raw water shows signs of high levels of toxic chemicals. The water is then sent to sand filtration beds to remove the remaining small particles, clarify the water, and enhance the effectiveness of disinfection. Chlorine, ozone, or ultraviolet light may be used as disinfectants.

At various points in the treatment process, the water is monitored, sampled, and tested using various physical, chemical, and microbial testing procedures. As the water leaves the treatment plant and enters the distribution system, chlorine is added as a disinfectant, particularly where ozone or ultraviolet light are used as disinfectants, to keep it free of microorganisms.

The water then goes to holding units, where it is stored until needed. These may be water towers, which use gravity to bring the water to the consumer without extra energy expense, or ground-level containers that require pumps to move the water. The water that ultimately flows from the tap should be clear, tasteless, and safe to drink.

Chlorination

The most extensively used disinfectant in the United States is chlorine, which is used to kill infectious microorganisms and parasites in water. Disinfection with chlorine or other similar chemicals prevents waterborne-disease outbreaks (WBDOs). The practice of chlorination began in the early 1900s to eliminate the cholera and typhoid outbreaks that were widespread in the United States.

In the early 1970s some scientific researchers became concerned by the possible health effects of total trihalo-methanes (TTHMs), a byproduct of chlorination. Chlorine reacts with naturally occurring organic substances in water to form TTHMs. The level of TTHMs formed varies widely across water supplies and is dependent on the amount of organic material in drinking water and the amount of chlorine applied. TTHMs are removed by passing the water through activated carbon filters.

The health effects of TTHMs are unclear. Some studies of human populations—such as Paolo Boffetta's "Human Cancer from Environmental Pollutants: The Epidemiological Evidence" (*Mutation Research*, September 28, 2006)—indicate a slightly higher incidence of bladder, lung, kidney, rectal, and colon cancer in areas where the water is chlorinated. Whereas other studies, such as Will D. King et al.'s "Case-Control Study of Colon and Rectal Cancers and Chlorination By-products in Treated Water" (*Cancer Epidemiology Biomarkers and Prevention*, 2000), report that an increased cancer risk is inconclusive.

Fluoridation

Fluoride, which occurs naturally in combination with other minerals in rocks and soils, is nature's cavity fighter. Water fluoridation is the process of adjusting the naturally occurring level of fluoride in most water

FIGURE 5.3

Water treatment process

Follow a drop of water from the source through the treatment process. Water may be treated differently in different communities depending on the quality of the water which enters the plant. Groundwater is located underground and typically requires less treatment than water from lakes, rivers, and streams.

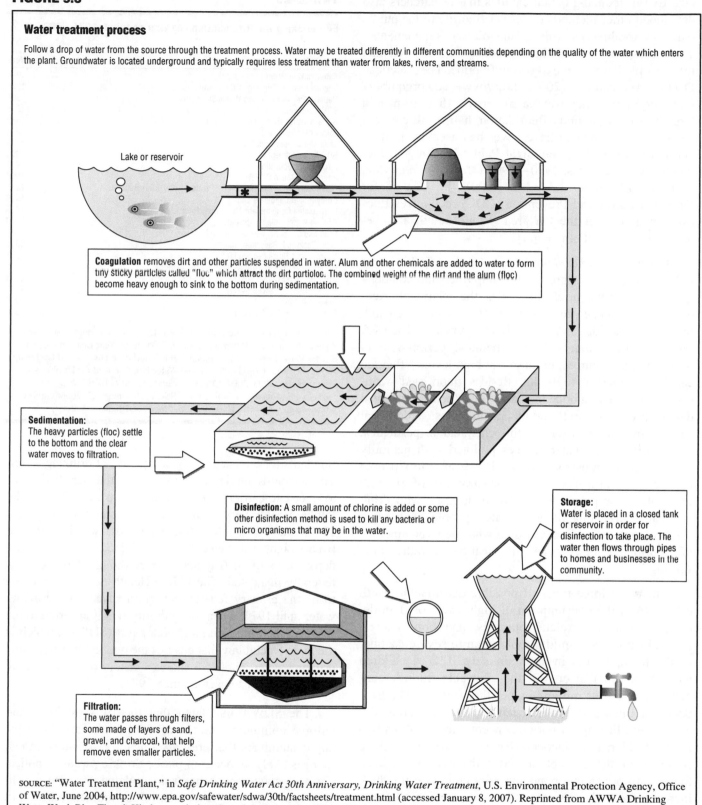

Lake or reservoir

Coagulation removes dirt and other particles suspended in water. Alum and other chemicals are added to water to form tiny sticky particles called "floc" which attract the dirt particles. The combined weight of the dirt and the alum (floc) become heavy enough to sink to the bottom during sedimentation.

Sedimentation:
The heavy particles (floc) settle to the bottom and the clear water moves to filtration.

Disinfection: A small amount of chlorine is added or some other disinfection method is used to kill any bacteria or micro organisms that may be in the water.

Storage:
Water is placed in a closed tank or reservoir in order for disinfection to take place. The water then flows through pipes to homes and businesses in the community.

Filtration:
The water passes through filters, some made of layers of sand, gravel, and charcoal, that help remove even smaller particles.

SOURCE: "Water Treatment Plant," in *Safe Drinking Water Act 30th Anniversary, Drinking Water Treatment*, U.S. Environmental Protection Agency, Office of Water, June 2004, http://www.epa.gov/safewater/sdwa/30th/factsheets/treatment.html (accessed January 8, 2007). Reprinted from AWWA Drinking Water Week Blue Thumb Kit, by permission. Copyright © American Water Works Association.

systems to a concentration (a range of 0.7 to 1.2 ppm) sufficient to protect against tooth decay. The decision to add fluoride to drinking water is left to each community. If the community elects to use fluoride, the water must meet the EPA maximum concentration limit.

In 1945 Grand Rapids, Michigan, became the first city in the world to add fluoride to its drinking water to prevent tooth decay. Since that time most community water systems in the United States have introduced water fluoridation. Fluoridation of drinking water proved so effective in reducing dental cavities that researchers also developed other methods to deliver fluoride to the public (such as toothpastes, rinses, and dietary supplements). The widespread use of these products has ensured that most people have been exposed to fluoride. The American Dental Association (2007, http://www.ada.org/public/manage/you/working_water.asp) reports that, thanks in large part to community fluoridation, half of all children aged five to seventeen have never had a cavity in their permanent teeth. In "Ten Great Public Health Achievements—United States, 1900–1999" (*Morbidity and Mortality Weekly Report*, April 2, 1999), the Centers for Disease Control and Prevention (CDC) recognizes fluoridation as one of the ten great public health achievements of the twentieth century.

In *Fluoride in Drinking Water: A Scientific Review of EPA's Standards* (March 2006, http://dels.nas.edu/dels/rpt_briefs/fluoride_brief_final.pdf), the National Academies of Science presents its most up-to-date research review on fluoridation and its effects on health. The EPA regulates the amount of this chemical permissible in drinking water sources, because—at high levels—fluoride can be detrimental to health. Besides fluoridated water, people can ingest fluoride in nonbeverage foods, fluoridated dental products, and certain medications. Additional sources include exposure to this chemical in pesticides, in the air, and in water sources polluted with naturally high concentrations of fluoride. However, most people exposed to harmfully high concentrations of fluoride are athletes or people with certain diseases who drink unusually large quantities of water, people exposed to fluoride in their work, or children who use large amounts of fluoridated tooth paste and swallow it rather than spitting it out.

There are three types of possible detrimental health effects from the consumption of high levels of fluoride: enamel fluorosis, skeletal fluorosis, and bone cancer. Enamel fluorosis, a mild discoloration to severe staining of the teeth, occurs in approximately 10% of children who drink water at or near the allowable fluoride concentration of four milligrams per liter (mg/L). The incidence of this condition is much lower in children who drink water having a fluoride concentration of about two mg/L. Evidence is inconclusive as to whether severe fluorosis of the enamel increases the risk of cavities. Skeletal fluorosis is a bone and joint condition in which increased bone density results in joint pain and stiffness. The incidence of skeletal fluorosis in the United States is rare. Research on the relationship of fluoridated water to bone cancer is inconclusive.

SAFE DRINKING WATER ACT

Federal regulation of drinking water quality began in 1914, when the U.S. Public Health Service established

TABLE 5.3

EPA drinking water regulations by year enacted, 1974–2006

Regulation	Year
Safe Drinking Water Act (SDWA)	1974
Interim Primary Drinking Water Standards	1975
National Primary Drinking Water Standards	1985
SDWA Amendments	1986
Surface Water Treatment Rule (SWTR)	1989
Total Coliform Rule	1989
Lead and Copper Regulations	1990
SDWA Amendments	1996
Information Collection Rule	1996
Interim Enhanced SWTR	1998
Disinfectants and Disinfection By-Products (D-DBPs) Regulation	1998
Contaminant Candidate List	1998
Unregulated Contaminant Monitoring Regulations	1999
Ground Water Rule (proposed)	2000
Lead and Copper Rule—action levels	2000
Filter Backwash Recycling Rule	2001
Long Term 1 Enhanced SWTR	2002
Unregulated Contaminant Monitoring Regulations	2002
Drinking Water Contaminant Candidate List 2	2005
Long Term 2 Enhanced SWTR	2006
Stage 2 D-DBP Rule	2006
Ground Water Rule finalized	2006

SOURCE: Jennifer L. Liang et al., "Table 1. U.S. Environmental Protection Agency Regulations Regarding Drinking Water, by Year Enacted—United States, 1974–2006," in "Surveillance for Waterborne Disease and Outbreaks Associated with Drinking Water and Water Not Intended for Drinking—United States, 2003–2004," *Morbidity and Mortality Weekly Report, Surveillance Summaries*, vol. 55, no. SS-12, December 22, 2006, http://www.cdc.gov/mmwr/pdf/ss/ss5512.pdf (accessed January 5, 2007)

standards for the bacteriological quality of drinking water. The standards applied only to systems that supplied water to interstate carriers such as trains, ships, and buses. The Public Health Service revised these standards in 1925, 1942, and 1962. The 1962 standards, which regulated twenty-eight substances, were adopted by the health departments of all fifty states, even though they were not federally mandated. The Public Health Service continued to be the primary federal agency involved with drinking water until 1974, when the authority was transferred to the EPA via the Safe Drinking Water Act (SDWA), which is the main federal law that ensures the quality of Americans' drinking water. Table 5.3 shows a history of EPA drinking water regulations enacted since 1974.

The SDWA mandated that the EPA establish and enforce minimum national drinking water standards for any contaminant that presents a health risk and is known to, or is likely to, occur in public drinking water supplies. For each contaminant that was regulated, the EPA was to set a legal limit on the amount of contaminant allowed in drinking water. In addition, the EPA was directed to develop guidance for water treatment and to establish testing, monitoring, and reporting requirements for water suppliers.

Congress intended that, after the EPA had set regulatory standards, each state would be granted primacy; that is, states would have the primary responsibility for enforcing the requirements of the SDWA. To be given primacy, a state must adopt drinking water standards and conduct monitoring and enforcement programs at least as stringent as those established by the EPA. Forty-nine states and all U.S. commonwealths and territories have received primacy. The EPA implements the drinking water program on Native American reservations. According to the fact sheet "Safe Drinking Water Act 30th Anniversary Drinking Water Monitoring, Compliance, and Enforcement" (June 2004, http://www.epa.gov/ogwdw/sdwa/30th/factsheets/monitoring_compliance.html), the EPA states that Wyoming and the District of Columbia have not received primacy over enforcement of the SDWA.

The EPA established the primary drinking water standards by setting MCLs for contaminants that are known to be detrimental to human health. The contaminants and their MCLs are shown in Table 4.1 in Chapter 4.

All public water systems in the United States are required to meet the primary standards. Only two contaminants regulated thus far, microorganisms and nitrates, pose an immediate health problem when the standards are exceeded. All other contaminants for which standards have been established must be controlled because ingesting water that exceeds these MCLs over a long period may cause long-term health problems, such as cancer, liver, or kidney disease or other harmful effects.

Secondary standards cover aspects of drinking water that have no health risks, such as odor, taste, staining properties, and color. Secondary standards are recommended but not required.

SDWA AMENDMENTS AND REGULATIONS

The EPA is continuing its work to protect drinking water from unsafe contaminant levels, to oversee the activities of the states that enforce federal or their own stricter standards, and to solicit public input as it develops new standards or other program requirements. Over the years the SDWA has been amended to require the EPA to:

- Set a maximum contaminant level goal (MCLG). An MCLG is the maximum amount of a contaminant that is not expected to cause any health problems over a lifetime of exposure. The EPA is mandated to set the MCL as close to the MCLG as technology and economics will permit.

- Specify the "best available technology" for treating each contaminant for which the EPA sets an MCL.

- Provide states with greater flexibility to implement the SDWA to meet their specific needs while arriving at the same level of public health protection.

- Set contaminant regulation priorities based on data about adverse public health effects of the contaminant, the occurrence of the contaminant in public water supplies, and the estimated reduction in health risk that can be expected from any new regulations.

- Provide a thorough analysis of the costs to water supplies and benefits to public health.

- Increase research to develop sound scientific data to provide a base for regulations.

- Ban the use of lead pipes and lead solder in new drinking water systems and in the repair of existing water systems.

- Establish a federal-state partnership for regulation enforcement.

COMPANION LEGISLATION TO THE SDWA
Water Efficiency Act of 1992

The Water Efficiency Act of 1992 established uniform national standards for manufacture of water-efficient plumbing fixtures, such as low-flow toilets and showers. The purpose was to promote water conservation by residential and commercial users.

According to the U.S. General Accounting Office (now the U.S. Government Accountability Office), in *Water Infrastructure: Water-Efficient Plumbing Fixtures Reduce Water Consumption and Wastewater Flows* (August 2000, http://www.gao.gov/new.items/rc00232.pdf), preliminary results from studies by the American Water Works Association and the EPA indicate that by 2020 water consumption could be reduced by 3% to 9% in the areas studied. Wastewater flows to sewage treatment plants could be reduced 13% by 2016. For the sixteen localities analyzed, the use of water-efficient plumbing fixtures could reduce the local water consumption enough to save local water utilities between $165.7 million and $231.2 million by 2020 because planned investments to expand drinking water treatment or storage capacity could be deferred or avoided.

Federal Water Pollution Control Act

The 1972 Federal Water Pollution Control Act (FWPCA) established the framework for regulating the discharge of pollutants to U.S. waters. This framework was strengthened by amendments in 1977 (the Clean Water Act) and in 1987 (the Water Quality Act).

The FWPCA and its amendments established the National Pollution Discharge Elimination System (NPDES) to reduce discharge of pollutants into water, including drinking water sources. States may apply for and receive primacy for the NPDES program in a manner similar to SDWA primacy.

The FWPCA also requires the EPA and the states to identify water resources that need to be cleaned up to meet water quality standards and to establish stringent controls where needed to achieve the water quality standards. States are required to develop lists of contaminated waters, to identify the sources and amounts of pollutants causing water quality problems, and to develop individual control strategies for the sources of pollution.

Aggressive use of the FWPCA by the EPA and the states can reduce the contaminant loads reaching drinking water sources. Preventing contaminants from reaching drinking water sources protects public health and reduces the need for and cost of water treatment instead of passing the costs on to the water consumer.

COST OF CLEAN DRINKING WATER

From 1976 to 2004 the number of contaminants regulated under the SDWA roughly quadrupled. (See Figure 5.4.) As a result, new treatment technologies have been required. This has significantly increased the cost of water treatment in many locations. More than ninety contaminants are now regulated.

Besides treating water, public water supply systems must:

• Protect their water source

• Build, maintain, and repair the treatment plants and distribution systems

• Replace aging systems

• Recruit, pay, and train system operation staff

• Meet the expanding treatment requirements of the SDWA and its monitoring and reporting requirements

• Expand service areas

• Provide necessary administrative and support services to accomplish these tasks

FIGURE 5.4

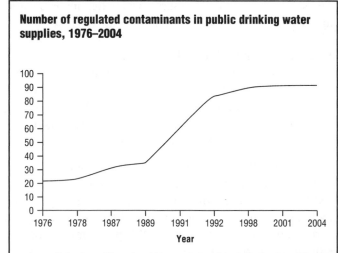

Number of regulated contaminants in public drinking water supplies, 1976–2004

SOURCE: "Number of Regulated Contaminants," in *Safe Drinking Water Act, Progress in Providing Safe Drinking Water*, U.S. Environmental Protection Agency, Office of Water, 2005, http://www.epa.gov/safewater/sdwa/30th/article_progress.html (accessed January 9, 2007)

Most of the money to support these services comes directly from users. The remainder of revenues comes from connection or inspection fees, fines, penalties, and other nonconsumption-based charges, as well as local or state grants or loans. The EPA, in *Water on Tap: What You Need to Know* (October 2003, http://www.epa.gov/safewater/wot/pdfs/book_waterontap_full.pdf), states that "despite rate increases, water is generally still a bargain compared to other utilities, such as electricity and phone service. In fact, in the United States, combined water and sewer bills average only about 0.5 percent of household income."

HOW CLEAN IS THE WATER?

Safe drinking water is a cornerstone of public health. Drinking water in the United States is generally safe. The vast majority of U.S. residents receive water from systems that have no reported violations of MCLs or flaws in treatment techniques, monitoring, or reporting.

However, in the Natural Resources Defense Council's 3,500-page report *Victorian Water Treatment Enters the 21st Century* (1994), Brian Cohen and Eric Olson document some 250,000 violations of the SDWA that occurred in 1991 and 1992. Cohen and Olson find that 43% of the water systems (serving about 120 million people) had committed violations.

Even though substantial progress has been made in reducing SDWA violations since this study was issued, the EPA estimates in *Factoids* that 79.1 million people (out of 282.3 million; see Table 5.2) were supplied with water from community water systems that registered one or more violations for health-based SDWA standards in 2004. (See Table 5.4.) In 2003 there were 81.7 million people affected by systems registering violations; this represents a decrease of 3% from 2003 to 2004.

However, the number of systems experiencing violations rose 4%, from 20,343 in 2003 to 21,200 in 2004. (See Table 5.5.) Furthermore, the number of violations increased 33%, from 88,695 in 2003 to 118,420 in 2004. (See Table 5.6.) The EPA suggests in *Factoids* that small

TABLE 5.4

Community water system violations by population affected, 2001–04

Fiscal year	Total
2004	79,058,089
2003	81,672,086
2002	56,644,512
2001	77,845,089

SOURCE: Adapted from "CWS Violations Reported by FY, Population Affected," in *Factoids: Drinking Water and Ground Water Statistics for 2005*, U.S. Environmental Protection Agency, Office of Water, December 2006, http://www.epa.gov/safewater/data/pdfs/statistics_data_factoids_2005.pdf (accessed January 8, 2007)

TABLE 5.5

Community water system violations by number of systems in violation, 2001–04

	Number of systems in violation
Fiscal year	**Total**
2004	21,200
2003	20,343
2002	20,232
2001	20,996

SOURCE: Adapted from "CWS Violations Reported by FY, Number of Systems in Violation," in *Factoids: Drinking Water and Ground Water Statistics for 2005*, U.S. Environmental Protection Agency, Office of Water, December 2006, http://www.epa.gov/safewater/data/pdfs/statistics_data_factoids_2005.pdf (accessed January 8, 2007)

TABLE 5.6

Community water system violations by number of violations, 2001–04

	Number of violations
Fiscal year	**Total**
2004	118,420
2003	88,695
2002	99,495
2001	82,655

SOURCE: Adapted from "CWS Violations Reported by FY, Number of Violations," in *Factoids: Drinking Water and Ground Water Statistics for 2005*, U.S. Environmental Protection Agency, Office of Water, December 2006, http://www.epa.gov/safewater/data/pdfs/statistics_data_factoids_2005.pdf (accessed January 8, 2007)

water systems are more likely to violate regulations than all other sizes. Because of their small size, manpower and funding to maintain and upgrade equipment and to meet monitoring and reporting requirements is extremely limited.

Disease Caused by Contaminated Drinking Water

It is difficult to know the exact incidence of illness caused by contaminated drinking water. People may not know the source of their illnesses and may attribute them to food poisoning, chronic illness, or infectious agents. Some researchers believe that the actual number of drinking water disease cases is higher than the reported number, but the diseases are not reported because victims believe them to be "stomach upsets" and treat themselves.

Since 1971 the CDC and the EPA have maintained a surveillance system for collecting and reporting data on WBDOs. In "Surveillance for Waterborne Disease and Outbreaks Associated with Drinking Water and Water Not Intended for Drinking—United States, 2003–2004" (*Morbidity and Mortality Weekly Report*, December 22, 2006), Jennifer L. Liang et al. examine CDC data about outbreaks associated with water intended for drinking

water and those associated with water used for recreation, such as beaches, hot tubs, and swimming pools.

During the 2003–04 period, eighteen states reported a total of thirty WBDOs in water intended for drinking—twelve in 2003 and eighteen in 2004. (See Figure 5.5.) Liang et al. report that those outbreaks caused an estimated 2,760 people to become ill and led to 4 deaths. As Figure 5.5 shows, five states had the highest number of outbreaks (three each) for the 2003–04 period: New York, New Jersey, Pennsylvania, Ohio, and Florida.

Table 5.7 lists the eleven WBDOs that were traced to contamination at or in the source water, treatment facility, or distribution system. They are listed by etiologic (disease-causing) agent and type of water system. Five of the outbreaks were associated with bacteria, one with parasites, two with chemicals, two with more than one microbe (mixed agents), and one unidentified. The highest percentage of outbreaks was associated with community (36%) and noncommunity (36%) water systems. However, more people were affected (72%) during the one outbreak with a mixed system. Liang et al. report that this mixed system involved noncommunity and individual water systems that both accessed sewage-contaminated groundwater. Cracks in the limestone aquifer allowed bacterial and viral contaminants to flow into the groundwater from the soil above, and people drank this untreated groundwater.

Table 5.8 lists the eight WBDOs that were traced to contamination at or in the source water or treatment facility. They are listed by etiologic agent and water source. Eighty-eight percent of the disease outbreaks and 97% of the cases (people affected) were associated with groundwater. The remainder were associated with surface water.

Figure 5.6 shows that WBDOs occurred year-round between 2003 and 2004 except for February. The number of outbreaks steadily increased during the spring and into July. The number dropped in August and leveled out through the fall, with an increase in November and another increase in January.

The number of WBDOs associated with drinking water in the United States declined from a peak of over fifty outbreaks in 1980. (See Figure 5.7.) Since 1987 there have been fewer than twenty WBDOs each year, except for 1992 and 2000.

Figure 5.8 shows the distribution of outbreaks during the 2003–04 period by disease-causing agents, water system, and water source. *Legionella* bacteria alone accounted for nearly one out of every four outbreaks (26.7%). (*Legionella* bacteria are commonly found in bodies of water, including air conditioning cooling towers, sink taps, and showerheads. Some species cause Legionnaires' disease, a pneumonia-like illness that often

FIGURE 5.5

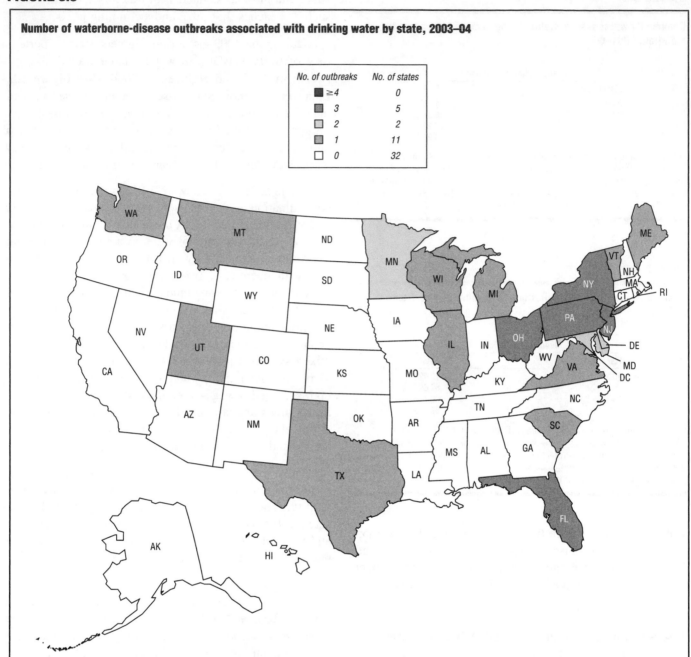

Number of waterborne-disease outbreaks associated with drinking water by state, 2003–04

No. of outbreaks	No. of states
≥4	0
3	5
2	2
1	11
0	32

Note: Sample size=30. Numbers are dependent on reporting and surveillance activities in individual states and do not necessarily indicate that more outbreaks occurred in a given state.

SOURCE: Jennifer L. Liang et al., "Figure 2. Number of Waterborne-Disease Outbreaks Associated With Drinking Water—United States, 2003–2004, in "Surveillance for Waterborne Disease and Outbreaks Associated with Drinking Water and Water Not Intended for Drinking—United States, 2003–2004," *Morbidity and Mortality Weekly Report, Surveillance Summaries*, vol. 55, no. SS-12, December 22, 2006, http://www.cdc.gov/mmwr/pdf/ss/ss5512.pdf (accessed January 5, 2007)

includes kidney, liver, and gastrointestinal symptoms as well.) All bacteria, including *Legionella*, were responsible for slightly more than 43% of outbreaks. Chemicals and other toxins accounted for another one-fourth (26.7%) of all outbreaks. Parasites and viruses were each the cause of 3.3% of outbreaks. Most (87.5%) of the outbreaks were linked to groundwater sources. Community and noncommunity water systems each experienced 36.4% of the outbreaks.

ARE AMERICANS CONCERNED ABOUT THEIR DRINKING WATER?

A Gallup poll conducted in 2006 indicates that water purity was one of the most important environmental concerns to Americans. Table 5.9 shows the results of this survey in which pollsters asked Americans to list their greatest environmental concerns along with their level of concern. Their greatest concerns focus on various aspects of water pollution. A majority (54%) had a great

TABLE 5.7

Number of waterborne-disease outbreaks (WBDOs) associated with drinking water, by causative agent and type water system, 2003–04

[Sample size=11]

Etiologic agent	Type of water system[a] Community WBDOs	Community Cases	Noncommunity WBDOs	Noncommunity Cases	Individual[b] WBDOs	Individual Cases	Mixed system WBDOs	Mixed system Cases	Total WBDOs	Total Cases
Bacteria	1	34	2	90	2	167	0	0	5	291
Campylobacter spp.	1	34	1	20	1	110	0	0	3	164
C.jejuni and shigella spp.	0	0	0	0	1	57	0	0	1	57
Salmonella typhimurium	0	0	1	70	0	0	0	0	1	70
Parasites	0	0	1	11	0	0	0	0	1	11
Giardia intestinalis	0	0	1	11	0	0	0	0	1	11
Chemicals/toxins	2	6	0	0	0	0	0	0	2	6
Sodium hydroxide	2	6	0	0	0	0	0	0	2	6
Mixed agents[c]	1	82	0	0	0	0	1	1,450	2	1,532
C.jejuni, C. lani, Cryptosporidium spp., and helicobacter canadensis	1	82	0	0	0	0	0	0	1	82
C.jejuni, norovirus, and G.intestinalis	0	0	0	0	0	0	1[d]	1,450	1	1,450
Unidentified	0	0	1	174	0	0	0	0	1	174
Unidentified	0	0	1	174	0	0	0	0	1	174
Total	**4**	**122**	**4**	**275**	**2**	**167**	**1**	**1,450**	**11**	**2,014**
Percentage	(36)	(6)	(36)	(14)	(18)	(8)	(9)	(72)	(100)	(100)

Note: WBDOs with deficiencies 1–4 (i.e., surface water contamination, groundwater contamination, water treatment deficiency, and distribution system contamination) were used for analysis.
[a]Community and noncommunity water systems are public water systems that have >15 service connections or serve an average of >25 residents for >60 days/year. A community water system serves year-round residents of a community, subdivision, or mobile home park. A noncommunity water system serves an institution, industry, camp, park, hotel, or business and can be nontransient or transient. Nontransient systems serve ≥25 of the same persons for ≥6 months of the year but not year-round (e.g., factories and schools), whereas transient systems provide water to places in which persons do not remain for long periods of time (e.g., restaurants, highway rest stations, and parks). Individual watersystems are small systems not owned or operated by a water utility that have <15 connections or serve <25 persons.
[b]Excludes commercially bottled water, therefore not comparable to previous summaries.
[c]Multiple etiologic agent types (e.g., bacteria, parasite, virus, and/or chemical/toxin) identified.
[d]Noncommunity and individual water systems.

SOURCE: Jennifer L. Liang et al., "Table 8. Number of Waterborne-Disease Outbreaks (WBDOs) Associated with Drinking Water (n=11), by Etiologic Agent and Type of Water System—United States, 2003–2004," in "Surveillance for Waterborne Disease and Outbreaks Associated with Drinking Water and Water Not Intended for Drinking—United States, 2003–2004," *Morbidity and Mortality Weekly Report, Surveillance Summaries*, vol. 55, no. SS-12, December 22, 2006, http://www.cdc.gov/mmwr/pdf/ss/ss5512.pdf (accessed January 5, 2007)

TABLE 5.8

Number of waterborne-disease outbreaks (WBDOs) associated with drinking water, by causative agent and type of water source, 2003–04

[Sample size=8]

Etiologic agent	Water source Ground water WBDOs	Ground water Cases	Surface water WBDOs	Surface water Cases	Unknown WBDOs	Unknown Cases	Mixed source WBDOs	Mixed source Cases	Total WBDOs	Total Cases
Bacteria	3	200	1	57	0	0	0	0	4	257
Campylobacter spp.	2	130	0	0	0	0	0	0	2	130
C. jejuni and shigella spp.	0	0	1	57	0	0	0	0	1	57
Salmonella typhimurium	1	70	0	0	0	0	0	0	1	70
Chemicals/toxins	2	6	0	0	0	0	0	0	2	6
Sodium hydroxide	2	6	0	0	0	0	0	0	2	6
Mixed agents*	1	1,450	0	0	0	0	0	0	1	1,450
C. jejuni, norovirus, and giardia intestinalis	1	1,450	0	0	0	0	0	0	1	1,450
Unidentified	1	174	0	0	0	0	0	0	1	174
Unidentified	1	174	0	0	0	0	0	0	1	174
Total	**7**	**1,830**	**1**	**57**	**0**	**0**	**0**	**0**	**8**	**1,887**
Percentage	(88)	(97)	(13)	(3)	0	0	0	0	(100)	(100)

Note: WBDOs with deficiencies 1–3 (i.e., surface water contamination, ground water contamination, and water treatment deficiency) were used for analysis.
*Multiple etiologic agent types (e.g., bacteria, parasite, virus, and/or chemical/toxin) identified.

SOURCE: Jennifer L. Liang et al., "Table 9. Number of Waterborne-Disease Outbreaks (WBDOs) Associated with Drinking Water (n=eight), by Etiologic Agent and Water Source—United States, 2003–2004," in "Surveillance for Waterborne Disease and Outbreaks Associated with Drinking Water and Water Not Intended for Drinking—United States, 2003–2004," *Morbidity and Mortality Weekly Report, Surveillance Summaries*, vol. 55, no. SS-12, December 22, 2006, http://www.cdc.gov/mmwr/pdf/ss/ss5512.pdf (accessed January 5, 2007)

FIGURE 5.6

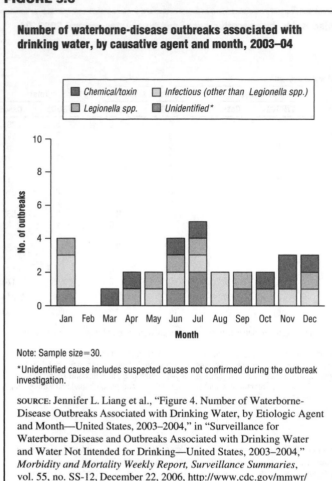

Number of waterborne-disease outbreaks associated with drinking water, by causative agent and month, 2003–04

Note: Sample size=30.

*Unidentified cause includes suspected causes not confirmed during the outbreak investigation.

SOURCE: Jennifer L. Liang et al., "Figure 4. Number of Waterborne-Disease Outbreaks Associated with Drinking Water, by Etiologic Agent and Month—United States, 2003–2004," in "Surveillance for Waterborne Disease and Outbreaks Associated with Drinking Water and Water Not Intended for Drinking—United States, 2003–2004," *Morbidity and Mortality Weekly Report, Surveillance Summaries*, vol. 55, no. SS-12, December 22, 2006, http://www.cdc.gov/mmwr/pdf/ss/ss5512.pdf (accessed January 5, 2007)

deal of concern about pollution of drinking water and 27% had a fair amount of concern. Thus, 81% were concerned about this issue. Respondents were also concerned about the contamination of the soil and water by toxic waste; pollution of rivers, lakes, and reservoirs; and maintenance of the nation's supply of freshwater for household needs.

Table 5.10 compares respondents' levels of concern regarding the pollution of drinking water with their levels of concern since 1990. Although a majority of Americans were concerned about this issue in 2006, the percentage of those who have a "great deal" of concern has dropped from a high of 72% in 2000. The percentages of those who care a "fair amount," "only a little," and "not at all" have risen since 1990.

BOTTLED WATER

Water is called bottled water only if it meets federal and state standards, is sealed in a sanitary container, and is sold for human consumption. The U.S. Food and Drug Administration (FDA) and state governmental agencies regulate bottled water as a packaged food product. The members of the International Bottled Water Association

(IBWA) produce and distribute about 85% of the bottled water sold in the United States. IBWA members must adhere to association standards besides those imposed by the government and undergo annual unannounced plant inspections by an independent third-party organization.

Imported European bottled water must meet the same federal and state standards. In addition, it must meet the strict standards set by the European Union. International bottler members who sell products in the United States must submit a certificate of inspection to the IBWA.

Growing Market

According to the Beverage Marketing Corporation in "2005 Stats" (2006, http://www.bottledwater.org/public/Stats_2005.doc), Americans consumed about 7.5 billion gallons of bottled water during 2005, a 10.7% increase from 2004. That figure translates into about twenty-six gallons per person in the United States, with bottled water ranking second only to carbonated soft drinks as the American beverage of choice.

Why Do Americans Like Bottled Water?

American consumers give a variety of reasons for their preference for bottled water. Some say they dislike the smell or taste of water from the tap or drawn from wells. Others cite the convenience of bottled water. In "Are Americans Financially Prepared for Disaster?" (Gallup Organization, October 18, 2005), Dennis Jacobe reports that 71% of Americans keep bottled water on hand in case of emergency.

Besides convenience, taste, and emergency use, concerns over the safety of public water supply systems since the September 11, 2001, terrorist attacks on the United States are prompting more Americans to rely on bottled water for their drinking water needs. According to the article "Take Me to the Water" (*Supermarket News*, May 26, 2003), a number of retailers report that concerns about terrorism have fueled a surge in the consumption of bottled water in the United States. The article states that between 1998 and 2003 sales of bottled water increased 150%.

DRINKING WATER SOURCE PROTECTION AND CONSERVATION

Because good sources of drinking water are a limited resource, the cost of developing and treating new sources is expected to rise. In addition, existing water suppliers are faced with the need to provide water to expanding service areas. As a result, the water industry is looking for cost-effective alternatives and is evaluating water conservation and reuse practices, as well as removing salt from seawater (desalinization) to create drinking water. Water suppliers are offering customers rebates

FIGURE 5.7

Number of waterborne-disease outbreaks associated with drinking water, by causative agent and year, 1971–2004

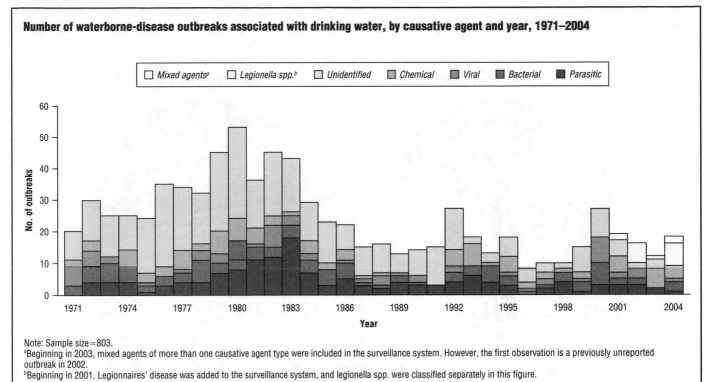

Note: Sample size=803.

[a]Beginning in 2003, mixed agents of more than one causative agent type were included in the surveillance system. However, the first observation is a previously unreported outbreak in 2002.

[b]Beginning in 2001, Legionnaires' disease was added to the surveillance system, and legionella spp. were classified separately in this figure.

SOURCE: Jennifer L. Liang et al., "Figure 3. Number of Waterborne-Disease Outbreaks Associated with Drinking Water, by Year and Etiologic Agent— United States, 1971–2004," in "Surveillance for Waterborne Disease and Outbreaks Associated with Drinking Water and Water Not Intended for Drinking—United States, 2003–2004," *Morbidity and Mortality Weekly Report, Surveillance Summaries*, vol. 55, no. SS-12, December 22, 2006, http://www.cdc.gov/mmwr/pdf/ss/ss5512.pdf (accessed January 5, 2007)

for using water-efficient toilets and showers, and in some areas are limiting the amount that can be used for lawn and landscape watering or car washing. In some locations municipal and county water departments are promoting the reuse of treated wastewater for irrigation and lawn watering instead of using precious drinking water.

WATER IN THE THIRD WORLD—A DEADLY DRINK

Water quality varies greatly in developing nations, as poverty often results in inadequate distribution of resources, including food and water, and sanitation practices are generally poor. According to the United Nations (UN), in *Water—A Shared Responsibility: World Water Development Report 2* (March 2006, http://unesdoc.unesco.org/images/0014/001444/144409E.pdf), one-sixth (1.1 billion) of the world's population is without access to an improved (safe and/or treated) water supply and 2.6 billion lack improved sanitation.

According to *The 1st UN World Water Development Report: Water for People, Water for Life* (March 2003, http://www.unesco.org/water/wwap/wwdr1/table_contents/index.shtml), as many as seven billion people in sixty countries could face water shortages by 2050. The report also suggests that pollution is a major problem, with 50% of the population in developing countries exposed to polluted water. A key goal, the UN report states, is to reduce by 50% the proportion of people who lack access to clean water by 2015.

Adequate quantities of safe water for drinking and for use in promoting personal hygiene are complementary measures for protecting public health. The lack of improved domestic water supply in the home leads to disease through two principal transmission routes: fecal-oral transmission and water-washed transmission.

In the fecal-oral transmission route, water contaminated with fecal material (sewage) is drunk without being treated or boiled, or food is prepared using this contaminated water, and waterborne disease occurs. Diseases transmitted by the fecal-oral route include typhoid, cholera, diarrhea, viral hepatitis A, dysentery, and dracunculiasis (guinea worm disease).

Water-washed transmission, the second route, is caused by a lack of sufficient quantities of clean water for washing and personal hygiene. People cannot keep their hands, bodies, and home environments clean and hygienic when there is not enough safe water available. The quantity of water that people use depends on their access to it. When water is available through a hose or house connection, people will use large quantities for hygiene. When water has to be hauled for more than a few minutes from source to home, the use drops

FIGURE 5.8

Percentage of waterborne-disease outbreaks associated with drinking water, by causative agent, water system, and water source, 2003–04

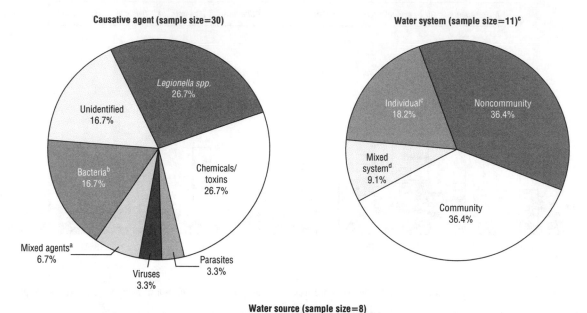

Causative agent (sample size=30)

Legionella spp. 26.7%

Unidentified 16.7%

Chemicals/toxins 26.7%

Bacteria[b] 16.7%

Mixed agents[a] 6.7%

Viruses 3.3%

Parasites 3.3%

Water system (sample size=11)[c]

Individual[c] 18.2%

Noncommunity 36.4%

Mixed system[d] 9.1%

Community 36.4%

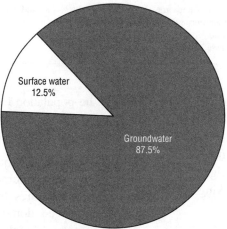

Water source (sample size=8)

Surface water 12.5%

Groundwater 87.5%

[a]Each waterborne-disease outbreak involves more than one causative agent.
[b]Other than *Legionella spp.*
[c]Does not include commercially bottled water, therefore, not comparable to previous summaries.
[d]Noncommunity and individual systems.

SOURCE: Jennifer L. Liang et al., "Figure 6. Percentage of Waterborne-Disease Outbreaks (WBDOs) Associated with Drinking Water, by Etiologic Agent, Water System, and Water Source—United States, 2003–2004," in "Surveillance for Waterborne Disease and Outbreaks Associated with Drinking Water and Water Not Intended for Drinking—United States, 2003–2004," *Morbidity and Mortality Weekly Report, Surveillance Summaries*, vol. 55, no. SS-12, December 22, 2006, http://www.cdc.gov/mmwr/pdf/ss/ss5512.pdf (accessed January 5, 2007).

significantly. Without enough clean water for good personal hygiene, skin and eye infections are easily spread, as are the fecal-oral transmission diseases.

The impact of poor water supply on human lives in developing and undeveloped countries is staggering. In *Ecosystems and Human Well-Being: Health Synthesis* (2005, http://www.who.int/globalchange/ecosystems/ecosysbegin.pdf), the

World Health Organization states these statistics: "Water-associated infectious diseases claim up to 3.2 million lives each year, approximately 6% of all deaths globally. The burden of disease from inadequate water, sanitation and hygiene totals 1.7 million deaths and the loss of more than 54 million healthy life years." Good drinking water, improved personal hygiene, and better sanitation practices would reduce this worldwide disease burden dramatically.

TABLE 5.9

Public opinion on concern about the environment, 2006

	Great deal	Fair amount	Great deal/ fair amount
	%	%	%
Pollution of drinking water	54	27	81
Contamination of soil and water by toxic waste	52	29	81
Pollution of rivers, lakes, and reservoirs	51	33	84
Maintenance of the nation's supply of fresh water for household needs	49	27	76
Air pollution	44	34	78
Damage to the Earth's ozone layer	40	28	68
The loss of tropical rain forests	40	24	64
The "greenhouse effect" or global warming	36	26	62
Extinction of plant and animal species	34	29	63
Acid rain	24	28	52

SOURCE: Joseph Carroll, "Environmental Concerns," in *Water Pollution Tops Americans' Environmental Concerns*, Gallup Poll News Service, April 21, 2006, http://www.galluppoll.com/content/?ci=22492&pg=1 (accessed February 1, 2007). Copyright © 2006 by The Gallup Organization. Reproduced by permission of The Gallup Organization.

TABLE 5.10

Public concern about pollution of drinking water, selected years 1990–2006

PLEASE TELL ME IF YOU PERSONALLY WORRY ABOUT THE POLLUTION OF DRINKING WATER A GREAT DEAL, A FAIR AMOUNT, ONLY A LITTLE, OR NOT AT ALL.

	Great deal	Fair amount	Only a little	Not at all	No opinion
	%	%	%	%	%
2006 Mar 13–16	54	27	12	7	*
2004 Mar 8–11	53	24	17	6	*
2003 Mar 3–5	54	25	15	6	—
2002 Mar 4–7	57	25	13	5	*
2001 Mar 5–7	64	24	9	3	*
2000 Apr 3–9	72	20	6	2	*
1999 Apr 13–14	68	22	7	3	*
1991 Apr 11–4	67	19	10	3	1
1990 Apr 5–8	65	22	9	4	*

*Less than 0.5% of respondents
—Question was not asked for a given profession in this survey.

SOURCE: "How much do you personally worry about ... pollution of drinking water?" in *Gallup's Pulse of Democracy: Environment*, Gallup Poll News Service, March 2006, http://www.galluppoll.com/content/default.aspx?ci=1615&pg=2 (accessed February 1, 2007). Copyright © 2006 by The Gallup Organization. Reproduced by permission of The Gallup Organization.

CHAPTER 6

OCEANS AND ESTUARIES

THE OCEAN

A view of the earth from a satellite shows an azure planet composed almost entirely of water. According to Tom S. Garrison, in *Oceanography: An Invitation to Marine Science* (2005), the ocean covers over two-thirds of the earth's surface to an average depth of 12,451 feet (or almost 2.3 miles). The U.S. Geological Survey (USGS), in *Where Is Earth's Water Located?* (August 28, 2006, http://ga.water.usgs.gov/edu/earthwherewater.html), notes that the earth contains 97% of the planet's water. It has a profound influence on the earth's environment. The ocean has subdivisions that are recognized as the Atlantic Ocean, the Pacific Ocean, the Indian Ocean, and the Arctic Ocean.

The terms *ocean* and *sea* are sometimes used interchangeably, but they are terms used out of tradition. According to Garrison, the ocean is "the vast body of saline water that occupies the depressions of the Earth's surface" and includes all the oceans and seas. For convenience sake, the ocean is subdivided into artificial compartments and given names that include the Atlantic, Pacific, and Indian oceans and the Mediterranean, Red, and Black seas.

Origin of the Ocean

Research suggests that the ocean is about four billion years old and that both the atmosphere on Earth and the ocean were formed through a process called outgassing of the earth's deep interior. According to this scientific theory, the ocean originated from the escape of water vapor from the melted rocks of the early Earth. The vapor rose to form clouds surrounding the cooling planet. After the earth's temperature had cooled to a point below the boiling point of water, rain began to fall and continued falling for millions of years. As this water drained into the huge hollows of the planet's cracked surface, the oceans were formed. The force of gravity kept this water on Earth. The oceans are still forming today at an extremely slow pace. According to

Garrison, about 0.025 cubic miles is added to the ocean each year.

Why Is the Ocean So Salty?

The salinity (saltiness) of the ocean is due to the presence of a high concentration of dissolved inorganic solids in water, primarily sodium and chloride (the components of table salt). Table 6.1 shows the principal constituents of ocean water. Early in the life of the planet, the ocean probably contained little of these substances. However, since the first rains descended on the young Earth billions of years ago and ran over the land, the rain has eroded the soil and rocks, dissolving them and transporting their inorganic solids to the ocean. Rivers and streams also carry dissolved inorganic solids and sediments and discharge them into the ocean. In *Why Is the Ocean Salty?* (1993, http://www.palomar.edu/oceanography/salty_ocean.htm), Herbert Swenson estimates that U.S. rivers and streams discharge 225 million tons of dissolved solids (salts) and 513 million tons of suspended sediment into the ocean each year. Throughout the world, rivers annually transport about four billion tons of dissolved salts to the ocean.

Over billions of years the ocean has become progressively more salty. The activity of the hydrological cycle concentrates the ocean salts as the sun's heat evaporates water from the surface of the ocean, leaving the salts behind. (See Figure 6.1.) There is so much salt in the ocean that if it could be taken out and spread evenly over the earth's entire land surface, it would form a layer more than five hundred feet thick—about the height of a forty-story building. (See Figure 6.2.)

Swenson notes that the salinity of the ocean is currently about thirty-five pounds per thousand pounds of ocean water, or thirty-five parts per thousand (ppt). This is similar to having a teaspoon of salt added to a glass of drinking water. By contrast, the U.S. Environmental Protection Agency (EPA), in *In Condition of the Mid-Atlantic Estuaries* (November

TABLE 6.1

Principal constituents of seawater

Chemical constituent	Content (parts per thousand)
Calcium (Ca)	0.419
Magnesium (Mg)	1.304
Sodium (Na)	10.710
Potassium (K)	0.390
Bicarbonate (HCO₃)	0.146
Sulfate (SO₄)	2.690
Chloride (Cl)	19.350
Bromide (Br)	0.070
Total dissolved solids (salinity)	**35.079**

SOURCE: Herbert Swenson, "Principal Constituents of Seawater," in *Why Is the Ocean Salty?* U.S. Geological Survey, 1993

FIGURE 6.1

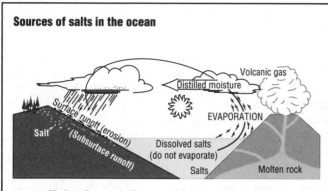

Sources of salts in the ocean

SOURCE: Herbert Swenson, "Sources of Salts in the Ocean," in *Why Is the Ocean Salty?* U.S. Geological Survey, 1993

FIGURE 6.2

The amount of salt in the ocean

If all the salt in the sea could be removed and spread over the Earth's surface, it would cover approximately one-half of the Empire State Building.

SOURCE: Herbert Swenson, "If All the Salt in the Sea Could be Removed and Spread over the Earth's Surface...," in *Why Is the Ocean Salty?* U.S. Geological Survey, 1993

1998, http://www.epa.gov/emap/html/pubs/docs/groupdocs/estuary/assess/cond_mae.pdf), indicates that freshwater has less than 0.5 ppt. Salinity in estuaries varies from slightly brackish (0.5 to 5 ppt) at the freshwater end to moderately brackish (5 to 18 ppt), to highly saline (19 ppt or more) near the ocean. An estuary is a body of water where a river meets the ocean and fresh- and saltwater mix.

Ocean as Controller of Earth's Climate

The ocean plays a major role in the earth's weather and long-term climate change. It has a huge capacity to store heat and can affect the concentration of atmospheric gases that control the planet's temperature. In *Environmental Health: Ecological Perspectives* (2005), Kathryn Hilgenkamp indicates that the top eight feet of the ocean hold as much heat as the entire atmosphere, making the ocean's ability to distribute heat an important factor in climate changes. For example, the occurrence of a southward-flowing current of warm water off the coast of western South America (El Niño), which is caused by a breakdown of trade wind circulation (steady winds blowing from east to west above and below the equator) can disrupt global weather patterns.

The ocean plays a crucial role in the cycle of carbon dioxide, a process affecting global warming. For example, the National Oceanic and Atmospheric Administration (NOAA) reports in "After Two Large Annual Gains, Rate of Atmospheric CO2 Increase Returns to Average, NOAA Reports" (*NOAA Magazine*, March 31, 2005), that the ocean stores some of the seven billion tons of carbon dioxide added each year to the atmosphere by natural sources and humankind's burning of fossil fuels. The ocean, trees, plants, and the soil serve as reservoirs for about half of all the human-produced carbon dioxide emitted each year since the Industrial Revolution, whereas the other half is accumulating in the atmosphere.

COASTAL POPULATIONS

According to the Population Reference Bureau (2006, http://www.prb.org/Content/NavigationMenu/PRB/Journalists/FAQ/Questions/Coastal_Population.htm), approximately 3 billion people worldwide, about half the world's population, live within 200 kilometers (about 125 miles) of a coastline. This living preference places huge segments of the world population at risk from coastal hazards, such as

TABLE 6.2

Leading states in coastal population growth, 1980–2003

State	Total change (million persons)	State	Percent change
California	9.9	Florida	75
Florida	7.1	Alaska	63
Texas	2.5	Washington	54
Washington	1.7	Texas	52
Virginia	1.6	Virginia	48
New York	1.6	California	47
New Jersey	1.2	New Hampshire	46
Maryland	1.2	Delaware	38
Michigan	0.8	Georgia	35
Massachusetts	0.7	South Carolina	33

SOURCE: Kristen M. Crossett et al., "Table 2. Leading States in Coastal Population Growth, 1980–2003," in *Population Trends Along the Coastal United States: 1980–2008*, National Oceanic and Atmospheric Administration, National Ocean Service, September 2004, http://marineeconomics.noaa.gov/socioeconomics/assessment/population/pdf/2_national_overview.pdf (accessed February 5, 2007)

TABLE 6.3

The deadliest mainland United States hurricanes, 1900–2006

Rank	Hurricane	Year	Category	Deaths
1	Unnamed (TX, Galveston)	1900	4	8,000[a]
2	Unnamed (FL, Lake Okeechobee)	1928	4	1,836
3	Katrina (Eastern LA, western MS)	2005	3	1,833
4	Unnamed (FL, Keys)	1919	4	600[b]
5	Unnamed (New England)	1938	3[c]	600
6	Unnamed (FL, Keys)	1935	5	408
7	Audrey (Southwest LA, inland TX)	1957	4	390
8	Unnamed (Northeast)	1944	3[c]	390[d]
9	Unnamed (LA, Grand Isle)	1909	4	350
10	Unnamed (LA, New Orleans)	1915	4	275
11	Unnamed (TX, Galveston)	1915	4	275

Note: Hurricanes, or tropical cyclones, were not named until the 1950s.
[a]This figure could be as high as 10,000 to 12,000.
[b]Over 500 lost on ships at sea, 600–900 estimated deaths.
[c]Moving more than 30 miles per hour.
[d]Of the total lost 344 were lost on ships at sea.

SOURCE: Adapted from Jerry D. Jarrel et al., "Table 2. The Thirty Deadliest Mainland United States Tropical Cyclones 1900–2000," in *The Deadliest, Costliest, and Most Intense United States Hurricanes from 1900 to 2000*, National Oceanic and Atmospheric Administration, National Weather Service, October 2001, http://www.aoml.noaa.gov/hrd/Landsea/deadly/Table2.htm (accessed January 9, 2007)

hurricanes, tidal waves, and flooding and increases pollution of both the ocean and estuaries.

In *Population Trends along the Coastal United States: 1980–2008* (September 2004, http://marineeconomics.noaa.gov/socioeconomics/assessment/population/pdf/1_front_matter_intro.pdf), Kristen M. Crossett et al. state that in 2003 an estimated 53% of the U.S. population (occupying only 26% of the total U.S. land mass) lived in coastal counties. From 1980 to 2003 the total coastal population of the United States increased by 27%, which is consistent with the nation's rate of increase as a whole. However, some coastal communities increased in population much more than others. Table 6.2 shows the leading states in coastal population growth, from 1980 to 2003. California had the greatest number of people move into the state, but Florida had the highest percentage growth.

Coastal Storms

The most common coastal hazard is the threat of the huge ocean storms that come ashore, generally during the warmer months of the year, and cause devastating damage to property and human life. These storms go by different names in different parts of the world. They are called hurricanes or tropical storms in the North Atlantic, the eastern North Pacific, and the western South Pacific. Typhoon is used for storms in the China Sea and the western North Pacific, whereas cyclone is used for storms in the Arabian Sea, the Bay of Bengal, and the South Indian Ocean.

In August 2005 Hurricane Katrina hit the U.S. Gulf Coast, causing widespread devastation in cities such as New Orleans, Louisiana; Mobile, Alabama; and Gulfport, Mississippi. Katrina was one of the strongest storms to reach the U.S. coast in the last 100 years with sustained winds during landfall of 125 miles per hour. Since 1900, it was the third deadliest hurricane (killing 1,833 people) and was by far the costliest hurricane. (See Table 6.3 and Table 6.4.)

Tsunamis are another coastal threat. A tsunami is an ocean wave—which may reach enormous dimensions—produced by a submarine (undersea) earthquake, landslide, or volcanic eruption. A December 2004 tsunami, triggered by a massive earthquake in the Indian Ocean, killed over two hundred thousand people and caused massive damage in Indonesia, Sri Lanka, India, Thailand, and many small islands in the region. The true death toll from the tsunami may never be known, and the devastation was so overwhelming that it is difficult to attach a dollar figure to it.

Much research is targeted at understanding and predicting coastal storms and tsunamis so that coastal residents can be warned of an impending event. Besides increasing the amount of property at risk, coastal population growth has created potentially life-threatening problems with storm warnings and evacuation. It has become increasingly difficult to ensure that the ever-rising numbers of residents and visitors can be evacuated and transported to adequate shelters during storm events. Sometimes hurricane evacuation decisions must be made well in advance of issuing hurricane warnings to mobilize the appropriate manpower and resources needed for the evacuation. Also, when a significant percentage of the coastal population has not experienced an event such as a hurricane, people are less likely to prepare and respond properly before, during, and after the event. However,

TABLE 6.4

The costliest U.S. hurricanes, 1900–2006

Rank	Hurricane	Year	Category	Damage in billions of dollars
1	Katrina (eastern LA, western MS coastlines)[a]	2005	3	$125
2	Andrew (southeast FL and LA)[a]	1992	5	36
3	Rita (TX-LA border coastlines)[a]	2005	3	16
4	Wilma (southwest FL)[a]	2005	3	16
5	Charley (FL, SC)[a]	2004	4	15
6	Ivan (AL)[a]	2004	3	14
7	Hugo (SC)[a]	1989	4	13.9
8	Frances (FL, GA, SC, NC, NY)[a]	2004	2	9
9	Jeanne (east central FL)[a]	2004	3	7
10	Agnes (FL, northeast U.S.)[b]	1972	1	8.6
11	Betsy (southeast FL and LA)[b]	1965	3	8.5
12	Camille (MS, southeast LA, VA)[b]	1969	5	7
13	Georges (FL Keys, MS, AL)[a]	1998	2	6.6
14	Floyd (mid atlantic & northeast U.S.)[a]	1999	2	6.5
15	Alicia (north TX)[a]	1983	3	5.9
16	Fran (NC)[a]	1996	3	5.8
17	Diane (northeast U.S.)[b]	1955	1	5.5
18	Isabel (NC, VA, MD)[a]	2003	2	5
19	Frederic (AL, MS)[b]	1979	3	5
20	Unnamed (New England)[b]	1938	3	4.7
21	Opal (northwest FL, AL)[a]	1995	3	3.6
22	Carol (northeast U.S.)[b]	1954	3	3.1
23	Juan (LA)[a]	1985	1	2.8
24	Carla (north & central TX)[b]	1961	4	2.6
25	Donna (FL, eastern U.S.)[b]	1960	4	2.4
26	Iniki (HI)[a]	1992	4	2.4
27	Elena (MS, AL, northwest FL)[a]	1985	3	2.4
28	Bob (NC, northeast U.S.)[a]	1991	2	2.1
29	Celia (south TX)[b]	1970	3	2
30	Dennis (FL, AL, GA, MS, TN)[a]	2005	3	2
31	Hazel (SC, NC)[b]	1954	4	1.9
32	Unnamed (FL, MS, AL)[b]	1926	4	1.7
33	Unnamed (north TX)[b]	1915	4	1.5
34	Dora (northeast FL)[b]	1964	2	1.5
35	Eloise (northwest FL)[b]	1975	3	1.5
36	Gloria (eastern U.S.)[b]	1985	3	1.5
37	Unnamed (northeast U.S.)[b]	1944	3	1.2
38	Beulah (south TX)[b]	1967	3	1.1
39	Bonnie (eastern NC & VA)[a]	1998	3	1.1

[a]Costs normalized to 2002 dollars using gross national product (GNP) Inflation/Wealth Index.
[b]Costs unadjusted for inflation.

SOURCE: Compiled in January 2007 by Sandra Alters for Thomson Gale, 2007, from Jerry D. Jarrel et al., "Table 3. Costliest U.S. Hurricanes 1900–2000 (unadjusted)," in *The Deadliest, Costliest, and Most Intense United States Hurricanes from 1900 to 2000*, National Oceanic and Atmospheric Administration, National Weather Service, October 2001, http://www.aoml.noaa.gov/hrd/Landsea/deadly/Table3.htm (accessed January 10, 2007) and "1980–2005 Billion Dollar U.S. Weather Disasters," in *Billion Dollar U.S. Weather Disasters*, National Oceanic and Atmospheric Administration, National Climatic Data Center, January 2007, http://www.ncdc.noaa.gov/img/reports/billion/disasterssince1980.pdf (accessed January 10, 2007)

following the 2004 tsunami, government officials and scientists began working to create a new tsunami warning system for that region.

CORAL REEFS—A SPECIAL OCEAN HABITAT

A coral reef is a submerged ridge near the surface of the water made up not only of colonies of coral animals that secrete hard skeletons but also of other aquatic organisms such as algae, mollusks, and worms. Coral reefs are among the richest marine ecosystems in terms of beauty, species, productivity, biomass (the amount of living matter), and structural complexity. They are dependent on intricate interactions between the coral, which provides the structural framework, and the organisms that live among the coral. Most reefs form as long narrow ribbons along the edge between shallow and deep waters, and their assets are many: fisheries for food, income from tourism and recreation, materials for new medicines, and shoreline protection from coastal storms.

Coral Reef Structure

Corals are simple, bottom-dwelling organisms related to the sea anemone and jellyfish. The basic building block of coral is a polyp, a tiny animal that has a common opening used to take in food and excrete wastes and is surrounded by a ring of tentacles. The weak stinging cells of the tentacles are used to capture small zooplankton (minute floating aquatic animals) for food. Each polyp sits in its own tiny bowl of calcium carbonate (its skeleton), which the coral constantly builds as it grows up from the ocean floor. Reef-building corals live in large colonies formed by the repeated divisions of genetically identical polyps, although many species of coral animals are represented within the reef. The colonies can take a wide variety of shapes, including branched, leafy, or massive forms, which may grow continuously for thousands of years.

Inside the sac of each coral polyp lives a single-celled algae. The algae give off oxygen and nutrients the coral uses to grow, and the coral gives off carbon dioxide and other substances the algae uses. Such a living situation is called symbiosis—the living together of two dissimilar organisms with mutual benefit.

Because of their dependence on symbiotic algae, coral reefs can grow only under conditions favoring the algae. Coral reefs are confined to tropical waters because the algae require warm, shallow, well-lit waters that are free of turbidity and pollution.

Coral Reefs—Ecosystems at Risk

The proximity of coral reefs to land makes them particularly vulnerable to the effects of human actions. Because they depend on light, coral reefs can be severely damaged by silt, which leads to an overgrowth of seaweed and other factors that reduce water clarity and quality. Sport diving and overfishing for food and the aquarium trade can deplete species and damage coral, resulting in disruption to the intricate interactions among reef species, as well as coral decline. (Overfishing means that so many fish are harvested that the natural breeding stock is depleted.) Introduction of exotic species through human activity can be devastating as the new predators consume the living reefs.

NOAA notes in "Major Reef-Building Coral Diseases" (January 23, 2007, http://www.coris.noaa.gov/about/diseases/) that along with becoming infected by bacteria or developing diseases of unknown causes, corals may respond to stress and damage with a condition known as coral bleaching. The coral expels the microscopic algae that normally live within its cells and provide the coral with their color, their ability to rapidly grow skeleton, and much of their food. The bleached coral turns pale, transparent, or unusual colors and then starves because it is unable to feed or reproduce. Increased bleaching is an early warning sign of deteriorating health and can be caused by extremes of light, temperature, or salinity.

Natural events, such as hurricanes, can also damage coral reefs. Healthy reefs generally recover from such damage, but unhealthy reefs often do not. In "Hazards to Coral Reefs" (January 23, 2007, http://www.coris.noaa.gov/about/hazards/), NOAA warns, "Current estimates note that 10 percent of all coral reefs are degraded beyond recovery. Thirty percent are in critical condition and may die within 10 to 20 years. Experts predict that if current pressures are allowed to continue unabated, 60 percent of the world's coral reefs may die completely by 2050."

Coral Reefs in the United States

The EPA notes in the *2000 National Water Quality Inventory* (August 2002, http://www.epa.gov/305b/2000report/chp4.pdf) that coral reefs are found in only three places in the United States: Florida (primarily in the Florida Keys), throughout the Hawaiian archipelago, and in the offshore Flower Gardens of Texas. The Florida reef system is part of the Caribbean reef system, the third largest barrier-reef ecosystem in the world. Five U.S. territories—American Samoa, Guam, the Northern Mariana Islands, Puerto Rico, and the U.S. Virgin Islands—also have lush reef areas. Figure 6.3 shows that the northwestern Hawaiian Islands make up 69% of the country's coral reef areas, by far the largest percentage in the United States and its territories.

Many U.S. coral reefs have been designated as marine sanctuaries with varying degrees of protection. The full extent and condition of most of these coral reefs is only beginning to be studied as a special area of focus.

In September 2002 NOAA released the first national assessment of the condition of coral reefs in the United States. The report, *The State of Coral Reef Ecosystems of the United States and Pacific Freely Associated States: 2002* (http://www.nccos.noaa.gov/documents/status_coralreef.pdf), was prepared under the auspices of the U.S. Coral Reef Task Force and established a baseline that is used for biennial reports on the health of coral reefs in the United States. NOAA also released the report *A National Coral Reef Action Strategy* (September 2002, http://www.coris.noaa

FIGURE 6.3

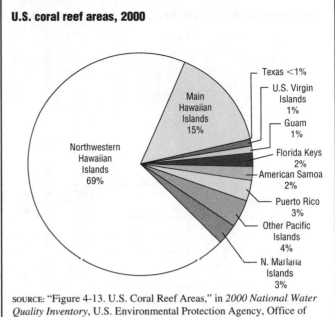

U.S. coral reef areas, 2000

SOURCE: "Figure 4-13. U.S. Coral Reef Areas," in *2000 National Water Quality Inventory*, U.S. Environmental Protection Agency, Office of Water, August 2002, http://www.epa.gov/305b/2000report/chp4.pdf (accessed January 10, 2007)

.gov/activities/actionstrategy/) to Congress outlining specific action to address thirteen major goals, including the continuation of mapping and monitoring, to protect coral reefs.

According to the *State of Coral Reef Ecosystems of the United States and Pacific Freely Associated States*, there are an estimated 7,607 miles of U.S. reefs and a range of 4,479 to 31,470 miles of reefs off the Freely Associated States. (The United States and Freely Associated States refer to fourteen jurisdictions that contain coral reef ecosystems, which are listed in Figure 6.4.) The report notes that an estimated 27% of the world's shallow water coral reefs may already be beyond recovery, and about 66% are severely degraded. The report also indicates that in all areas some coral reefs in the United States are in good to excellent health. However, every reef system is suffering from both human and natural disturbances. These reefs suffer from the same problems as do reefs all over the world, especially those resulting from rapidly growing coastal populations. The report states that 10.5 million people now live in U.S. coastal areas next to shallow coral reefs, and every year about 45 million people visit the areas.

Florida and the U.S. Caribbean were considered to be in the most unfavorable condition, mainly because of nearby dense populations and the effects of hurricanes, disease, overfishing, and a proliferation of algae. The NOAA report indicates that live coral cover in the Florida Keys has declined 37% since 1997. Since 1982, white-band disease has killed nearly all the elkhorn and staghorn

FIGURE 6.4

Comparisons of threat levels to coral reef ecosystems in the U.S. and Freely Associated States (FAS), 2002–04

		Climate change & coral bleaching	Diseases	Tropical storms	Coastal development and runoff	Coastal pollution	Tourism and recreation	Fishing	Trade in coral and live reef species	Ships, boats, and groundings	Marine debris	Aquatic invasive species	Security training activities	Offshore oil and gas exploration	Other	Jurisdictional composite trend	Change (2002 to 2004)
U.S. Virgin Islands	2002	✗	✔	✔	✔	✔	✗	✔	O	✔	O	O	O	O	O	14	↑
	2004	✗	✔	✔	✔	✔	✔	✔	O	✔	✗	O	O	O	O	16	
Puerto Rico	2002	✗	✔	O	✔	✔	✗	✔	✔	✗	✗	O	✔	O	✔	18	↓
	2004	✗	✗	✗	✔	✗	✗	✔	✗	✗	O	O	O	O	O	11	
Navassa	2002																N/A
	2004	O	✗	✗	O	O	O	✔	O	O	O	O	O	O	O	4	
Florida	2002	✔	✔	✗	✔	✔	✗	✔	✗	✔	✗	✗	O	O	O	17	↑
	2004	✔	✔	✗	✔	✔	✗	✔	✗	✔	✔	✗	O	O	✗	18	
Flower Gardens Banks	2002	O	O	O	O	O	O	O	O	✗	O	✗	O	✗	O	3	↑
	2004	O	O	O	O	O	O	✗	O	✗	O	✗	O	✗	O	4	
Main Hawaiian Islands	2002	O	O	O	✔	✔	✔	✔	✔	✔	✗	✔	✗	O	✗	17	-
	2004	✗	✗	O	✔	✔	✔	✔	✔	✗	✗	✔	✗	O	O	17	
Northwestern Hawaiian Islands	2002	O	O	O	O	O	✗	✗	✗	✔	✔	✔	O	O	O	9	↓
	2004	✗	O	O	O	O	O	✗	O	✗	✔	O	O	O	O	5	
American Samoa	2002	✗	O	✗	✔	✔	O	✔	✗	✗	O	✗	O	O	O	11	↓
	2004	✔	✗	✗	✔	✗	O	✔	O	O	O	O	O	O	O^a	9	
Pacific remote island areas	2002	✗	O	O	O	O	O	✗	O	✗	✗	✗	O	O	✔	7	↓
	2004	✗	O	O	O	O	O	✗	O	✗	✗	O	O	O	✗	5	
Marshall Islands	2002	✔	O	✗	✗	O	O	✗	✔	O	O	✗	O	O	O	8	↓
	2004	✔	O	✗	✗	O	O	✗	✗	O	O	✗	O	O	O	7	
Federated States of Micronesia	2002	✗	O	O	✔	✗	O	✔	O	✔	✔	O	O	O	✗	11	↓
	2004	O	O	O	✔	O	O	✔	O	✔	O	O	O	O	O	6	
Commonwealth of Northern Mariana Islands (CNMI)^b	2002	✗	O	✗	✔	✔	✗	✗	O	✗	✗	O	✔	O	✔	14	↑
	2004	✗	O	✗	✗	✗	O	✗	O	✗	O	O	✗	O	✔	9	
Guam	2002	O	O	✗	✔	✔	✗	✗	O	✗	O	O	O	O	O	8	↑
	2004	✔	O	✗	✔	✔	✔	✔	O	✗	O	O	O	O	O	13	
Palau	2002	✔	O	O	✔	✔	✗	✗	O	✗	✗	✗	O	O	O	11	↓
	2004	✔	O	O	✔	O	✗	✗	O	O	O	O	O	O	O	6	
Stressor change assessment	2002	12	6	7	19	17	9	18	9	17	10	10	5	1	8		
	2004	16	7	8	18	11	9	20	5	13	7	5	2	1	4		
Change (2002 to 2004)		↑	↑	↑	↓	↓	—	↑	↓	↓	↓	↓	↓	—	↓		
Temporal composite threat		✔	O	✗	✔	✔	✗	✔	O	✔	✗	✗	O	O	O		

Notes: ✔ represents high threat (2 points), ✗ represents moderate threat (1 point), and O represents little or no threat (0 points). Scores were tallied horizontally to calculate the level of threat from individual stressors across jurisdictions and vertically to calculate overall threat by jurisdiction for all stressors combined. Horizontal bars indicate no change. Only data for 2004 are available for Navassa. The actual impacts of each threat category will likely vary widely within and among regions.
^aFollowing the 2000 census, population growth emerged as a major issue in American Samoa.
^bFor CNMI, 2002 data were based on the southern islands only, while 2004 data include the northern islands; the perceived threat for the southern islands did not change from 2002 to 2004.

SOURCE: Jenny E. Waddell, editor, "Table 18.3. A Comparison of the 2002 and 2004 Perceived Levels of Threat to Coral Reef Ecosystems in the U.S. and FAS, Based on Expert Opinion," in *The State of Coral Reef Ecosystems of the United States and Pacific Freely Associated States: 2005*, National Oceanic and Atmospheric Administration, National Centers for Coastal Ocean Science, Center for Coastal Monitoring and Assessment's Biogeography Team, 2005, http://ccma.nos.noaa.gov/ecosystems/coralreef/coral_report_2005/NationalSummary_Ch18_C.pdf (accessed February 5, 2007)

corals off the coasts of St. Croix (U.S. Virgin Islands), Puerto Rico, and southeastern Florida.

In August 2005 NOAA released the second national assessment of the condition of coral reefs in the United States, *The State of Coral Reef Ecosystems of the United States and Pacific Freely Associated States: 2005*, (http://ccma.nos.noaa.gov/ecosystems/coralreef/coral_report_2005/CoralReport2005_C.pdf), which was edited by Jenny E. Waddell. Waddell notes progress in building a mapping and monitoring system. She also indicates that many local action strategies have been developed to counter threats to coral reefs, especially in priority threat areas. In separate chapters, each reporting area describes in quantitative detail the state of the coral reefs in their area. This differs from the more general qualitative (descriptive) nature of the first report.

A summary of a comparison between the 2002 and 2004 perceived levels of threat to coral reef ecosystems is shown in Figure 6.4. The rows compare each jurisdiction for all threat categories for 2002 and 2004. In Florida, for example, the threat level in each threat category remained the same from 2002 to 2004, except for the "other" category. Thus, the composite trend rose by one point from 2002 to 2004, as shown in the far right column, resulting in a increase of threat overall (arrow).

The columns compare each threat category for all jurisdictions for 2002 and 2004. For example, climate change, diseases, tropical storms, and fishing increased in threat level over those two years across all jurisdictions. However, eight of fourteen threat categories were perceived to have decreased in severity since 2002.

NEARSHORE WATERS

Nearshore waters are shallow waters a short distance from the shore in lakes, rivers, estuaries, and the ocean. Depending on the size of the water body, the nearshore waters may be minimal in size (a small lake) or large (the coastal waters of the Atlantic Ocean). They reflect the conditions and activities within the watershed. A watershed is an area in which water, sediments, and dissolved materials drain to a common outlet, such as a lake, river, estuary, or the ocean.

Whether marine, estuarine, or fresh, nearshore waters serve a variety of functions. They are the prime recreational waters, providing opportunities for swimming, boating, diving, surfing, snorkeling, and fishing. Nearshore waters are intimately linked with wetlands and sea grasses and provide a unique habitat for a variety of plants and animals. According to the EPA, in "Nearshore Waters and Your Coastal Watershed" (July 1998, http://www.epa.gov/owow/oceans/factsheets/fact3.html), these waters are the source of food and shelter for many species of fish and shellfish and provide habitat for 80% of the

fish species in the United States. Nearshore waters also provide many opportunities for education and research for students, naturalists, and scientists.

Because of their proximity to the shoreline, nearshore waters are particularly vulnerable to pollution. As a result, water quality in most confined waters and some nearshore waters is deteriorating, which in turn affects the plant and animal life. Besides pollution, nearshore waters are vulnerable to the everyday (and to all appearances, harmless) activities of people. For example, swimming has been restricted in some shallow lagoons with coral reefs and beautiful beaches because heavy use by swimmers resulted in chemical concentrations of suntan oil and sunblock lotion in the water that was high enough to kill or impair the coral reefs. Wakes from recreational powerboats in high-use areas have been shown to increase wave action, resulting in increased shoreline erosion. Increased pollutant levels from boat paints, spills during refueling, and leaks of gas and oil from recreational boat engines in areas of high recreational use affect both plants and animals. Private pier and boathouse construction result in shading of water, which contributes to sea grass decline. Balancing the need to accommodate the public's desire to enjoy water-related activities and ownership of waterfront property and the need to protect nearshore waters is a difficult management issue.

Estuaries

Estuaries are places of transition, where rivers meet the sea. An estuary is a partially enclosed body of water formed where freshwater from rivers flows into the ocean, mixing with the salty seawater. As mentioned earlier, salinity in estuaries varies from slightly brackish (0.5 to 5 ppt) at the freshwater end to moderately brackish (5 to 18 ppt), to highly saline (19 ppt or more) near the ocean. Although influenced by the tides, estuaries are protected from the full force of ocean waves, winds, and storms by reefs, barrier islands, or fingers of land, mud, or sand that make up their seaward boundary. Estuaries come in all shapes and sizes. Examples include the Chesapeake Bay, Puget Sound, Boston Harbor, San Francisco Bay, and Tampa Bay. In "The National Estuary Program: A Ten Year Perspective" (October 4, 2006, http://www.epa.gov/owow/estuaries/aniv.htm), the EPA states that there are approximately 130 estuaries in the United States.

The tidal sheltered waters of estuaries support unique communities of plants and animals that are specially adapted for life under a wide range of conditions. Estuarine environments are incredibly productive, producing more organic matter annually than any equal-sized area of forest (including rain forests), grassland, or cropland. A wide range of habitats exists around and in estuaries, including shallow open water, tidal pools, sandy beaches, mud and sand flats, freshwater and salt marshes, rocky shores, oyster reefs, mangrove forests, river deltas, wooded swamps, and kelp and sea grass beds.

NATIONAL ESTUARY PROGRAM. The Water Quality Act of 1987 created the National Estuary Program (NEP) to help achieve long-term protection of living resources and water quality (the basic "fishable/swimmable" goal of the Clean Water Act) in estuaries. To improve an estuary, the NEP brings together community members to define program goals and objectives, identify estuary problems, and design action plans to prevent or control pollution, while restoring habitats and living resources such as shellfish. This approach results in the adoption of a comprehensive conservation and management plan for implementation in each estuary. This integrated watershed-based, stakeholder-oriented, water resource management approach has led to some significant local environmental improvements since its founding. The EPA notes in "National Estuary Program" that in 1987 the NEP consisted of six estuary programs located throughout the nation; in 2006 there were twenty-eight estuary programs in eighteen states and Puerto Rico.

In "National Estuary Program Success Stories" (December 12, 2006, http://www.epa.gov/owow/estuaries/success.htm), the EPA mentions several NEP successes. Two examples of environmental improvement resulting from the NEP can be found in the Leffis Key and Corpus Christi projects. The Leffis Key restoration project in Sarasota Bay, Florida, resulted in thirty acres of productive intertidal habitat being created and planted with more than fifty thousand native plants and trees at a cost of $315,000. In Corpus Christi Bay, Texas, treated biosolids were applied to a twenty-five-acre plot of aluminum mine tailings, resulting in plant growth promotion, wildlife habitat, and improved water quality. Biosolids are composed of sewage sludge that has been properly treated and processed to make a nutrient-rich material that can be safely recycled and applied as fertilizer.

CHESAPEAKE BAY PROGRAM. According to the Chesapeake Bay Program Office (August 19, 2003, http://www.chesapeakebay.net/info/bayfaq.cfm#big), the Chesapeake Bay (the Bay) is the largest estuary in North America and one of the most productive estuaries in the world. It has a sixty-four-thousand-square-mile watershed that encompasses six states and the District of Columbia. Its watershed is home to more than fifteen million people and thirty-six hundred species of plants and animals. The Bay has over 11,600 miles of shoreline and averages 21 feet deep, with hundreds of thousands of acres of shallow water. It is two hundred miles long and thirty-five miles wide at its widest point.

The first estuary in the United States to be targeted for restoration and protection, the Bay is protected under its own federally mandated program, separate from the

NEP. The Chesapeake Bay Program (2007, http://www.chesapeakebay.net/) began in 1983 with a meeting of the governors of Maryland, Pennsylvania, and Virginia; the mayor of the District of Columbia; and the EPA administrator. These individuals signed the Chesapeake Bay Agreement committing their states and the District of Columbia to prepare plans for protecting and improving water quality and living resources in the Chesapeake Bay. The Chesapeake Bay Program evolved as the institutional mechanism to restore the Bay and to meet the goals of the Chesapeake Bay Agreement.

Although great progress has been made in the Chesapeake Bay restoration, much remained to be done by early 2007. The editorial "A Cleaner Bay, a Step at a Time" (*Salisbury Daily Times*, February 8, 2007) notes that increased population growth in the Chesapeake Bay area is contributing to ongoing problems in the estuary. The editorial sums up the work still needed to meet a 2010 cleanup deadline:

- About 1 million failing septic tanks across the regional watershed need to be dug up and repaired.

- Farmers—some 80,000 of them—would have to make expensive and radical changes to the way they manage their land and their fertilizer application.

- Hundreds of municipal and private sewage plants would have to be overhauled, costing millions of dollars each.

- In all, about $28 billion would have to be spent, or double the $14 billion already being devoted toward the restoration effort.

CONDITION OF THE NATION'S ESTUARIES. The EPA, NOAA, the USGS, the U.S. Fish and Wildlife Service, coastal states, and the National Estuary Programs coordinate efforts to produce the *National Coastal Condition Report*, which describes the ecological and environmental conditions in U.S. coastal waters, including estuaries. The most recent report is the *National Coastal Condition Report II (2005)* (December 2004, http://www.epa.gov/owow/oceans/nccr/2005/downloads.html). Figure 6.5 is a summary graphic from this report and shows the overall national coastal condition for estuaries.

The report uses five indicators of estuarine condition. (See Figure 6.5.) For each indicator, researchers assessed the condition at 2,073 estuary sites in the "lower 48" states between 1997 and 2000. The second step was to assign a regional rating. The water quality index consists of indicators such as nitrogen and phosphorus levels, water clarity, and level of dissolved oxygen. The sediment quality index refers to the level of contamination of the sediment with toxic chemicals. The benthic index refers to organisms that live at the bottom of estuaries. A good benthic index is one in which a wide variety of

FIGURE 6.5

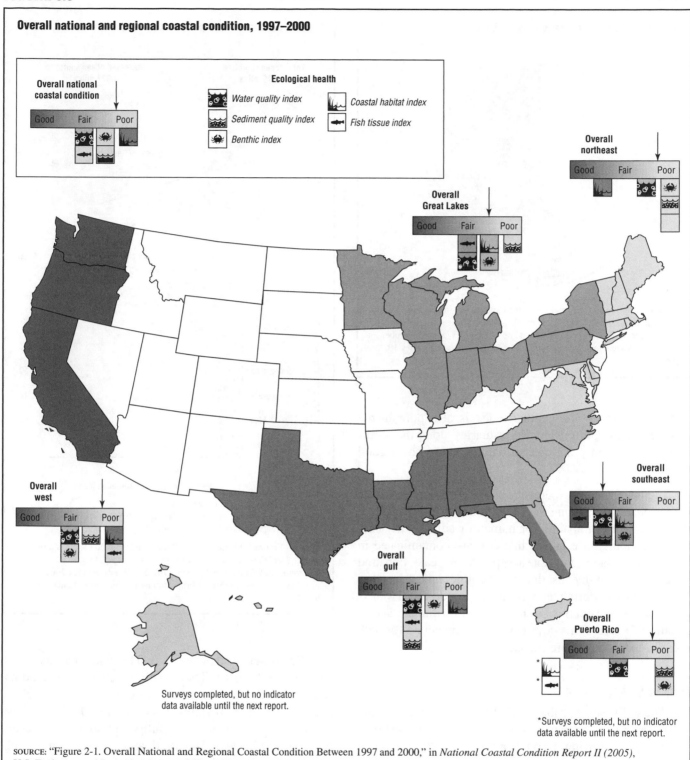

Overall national and regional coastal condition, 1997–2000

SOURCE: "Figure 2-1. Overall National and Regional Coastal Condition Between 1997 and 2000," in *National Coastal Condition Report II (2005)*, U.S. Environmental Protection Agency, Office of Research and Development and Office of Water, December 2004, http://www.epa.gov/owow/oceans/nccr/2005/Chap2_National_A.pdf (accessed January 10, 2007)

benthic species are found, of which there are few pollution-tolerant species and several pollution-sensitive species that are found. The coastal habitat index is an assessment of the loss of wetland areas (the terrestrial-aquatic interface) of estuarine ecosystems. The fish tissue index refers to levels of chemical contaminants within fish.

The *National Coastal Condition Report II* states that "the overall condition of estuaries in the United States is fair." The report also notes that estuaries in the Northeast have poorer water quality conditions than those in other regions of the country. The sediment quality index is poor in the estuaries of the Great Lakes, the Northeast,

FIGURE 6.6

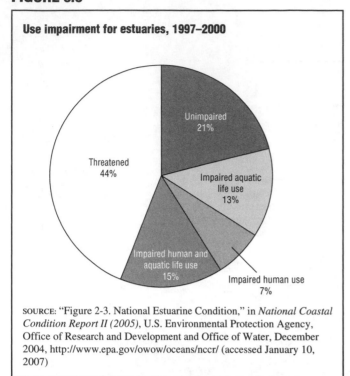

Use impairment for estuaries, 1997–2000

SOURCE: "Figure 2-3. National Estuarine Condition," in *National Coastal Condition Report II (2005)*, U.S. Environmental Protection Agency, Office of Research and Development and Office of Water, December 2004, http://www.epa.gov/owow/oceans/nccr/ (accessed January 10, 2007)

FIGURE 6.7

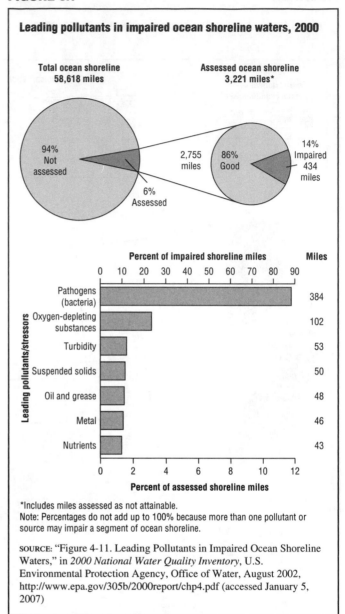

Leading pollutants in impaired ocean shoreline waters, 2000

*Includes miles assessed as not attainable.
Note: Percentages do not add up to 100% because more than one pollutant or source may impair a segment of ocean shoreline.

SOURCE: "Figure 4-11. Leading Pollutants in Impaired Ocean Shoreline Waters," in *2000 National Water Quality Inventory*, U.S. Environmental Protection Agency, Office of Water, August 2002, http://www.epa.gov/305b/2000report/chp4.pdf (accessed January 5, 2007)

and Puerto Rico. The benthic index is also poor in the Northeast and Puerto Rico. For estuaries along the northeastern coast, only the coastal habitat index has a good rating.

Figure 6.6 shows percentages of estuary area in the United States (excluding the Great Lakes) that is impaired, threatened, or unimpaired for human or aquatic life uses. Impaired human use means that fish are contaminated in those waters and are not acceptable to eat. Impaired aquatic life use means that the benthic index is poor. Threatened use correlates with a fair condition. Twenty-one percent of estuaries are unimpaired for human and/or aquatic life uses. Forty-four percent is threatened for both uses. Fifteen percent are impaired for both uses.

U.S. WATERWAYS

The EPA, states, tribes, and other federal agencies are collaborating on a new process to monitor the nation's waterways. Following the publication of the *2000 National Water Quality Inventory*, the EPA entered a transition period in the gathering and analysis of water quality data in nationally consistent, statistically valid assessment reports. The EPA article "Schedule for Statistically Valid Surveys of the Nation's Waters" (December 5, 2005, http://www.epa.gov/owow/monitoring/guide.pdf) details its new reporting schedule.

In the *2000 National Water Quality Inventory*, the EPA reports that fourteen of the twenty-seven coastal states and territories had rated the water quality of some

of their coastal waters in 2000. The states had assessed 14% of the 22,618 miles of national coastline excluding Alaska, or 5.5% (3,221 miles) of ocean shoreline (including Alaska's 36,000 miles of coastline). Of the 14% of ocean waters assessed, 79% fully supported their designated uses, 14% were impaired, and 7% were supporting uses but threatened. Designated uses (such as fishing and drinking water supply) are the beneficial water uses assigned to each water body by a state as part of its water quality standards.

The EPA reports in the *2000 National Water Quality Inventory* that bacteria (pathogens, which are disease-causing organisms) were identified as the leading contaminants of ocean shoreline waters, followed by oxygen-depleting substances and turbidity (cloudiness) in 2000. (See Figure 6.7.) Bacteria provide evidence of possible fecal contamination

FIGURE 6.8

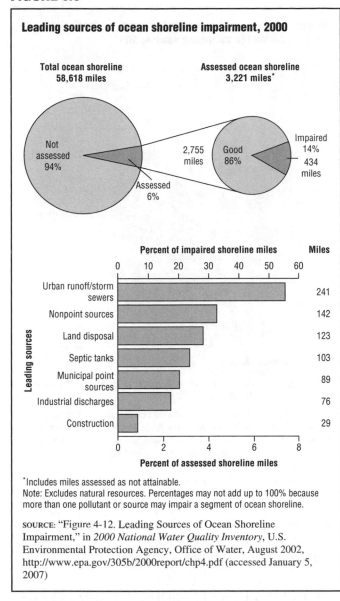

Leading sources of ocean shoreline impairment, 2000

Total ocean shoreline
58,618 miles

Assessed ocean shoreline
3,221 miles*

Not assessed 94%

Assessed 6%

2,755 miles

Good 86%

Impaired 14% 434 miles

Leading sources	Miles
Urban runoff/storm sewers	241
Nonpoint sources	142
Land disposal	123
Septic tanks	103
Municipal point sources	89
Industrial discharges	76
Construction	29

Percent of impaired shoreline miles

Percent of assessed shoreline miles

*Includes miles assessed as not attainable.
Note: Excludes natural resources. Percentages may not add up to 100% because more than one pollutant or source may impair a segment of ocean shoreline.

SOURCE: "Figure 4-12. Leading Sources of Ocean Shoreline Impairment," in *2000 National Water Quality Inventory*, U.S. Environmental Protection Agency, Office of Water, August 2002, http://www.epa.gov/305b/2000report/chp4.pdf (accessed January 5, 2007)

bottom-dwelling organisms such as oysters. Turbidity was responsible for more than 10% of the impaired ocean shoreline miles reported to the EPA in 2000. (See Figure 6.7.) Three of the leading sources of ocean impairment are also contributors to turbidity: runoff from highly developed urban areas, agricultural activities (nonpoint sources), and construction projects. (See Figure 6.8.)

The EPA reports in the *2000 National Water Quality Inventory* that most of the ocean waters assessed supported the five general-use categories shown for estuaries: aquatic life support, fish consumption, shellfishing, primary contact, and secondary contact. These categories represent summaries of the designated uses and their achievement provided by the states to the EPA. Waters that either support their designated uses only part of the time or do not support their uses at all are considered impaired. Good water quality supports primary contact (swimming without risk to public health) in 85% of the assessed ocean waters (the same percentage for use support in estuaries) and fish consumption (fish safe to eat) in 91% (compared with 52% of use support in estuaries).

In 94% of the waters assessed, the water was considered of good quality and capable of supporting aquatic life (suitable habitat for protection and propagation of desirable fish, shellfish, and other aquatic organisms). In the shellfish harvesting summary (water quality supports a population of bivalves free from toxicants and pathogens that can pose a health risk to people who eat them), 86% of the ocean waters assessed had good water quality that supported this use. In addition, good water quality in 91% of the ocean waters assessed supported secondary contact recreation (people can perform water-based activities such as waterskiing and boating without risk of adverse human health effects).

BEACHES

Beach closings take place hundreds of times each year to protect the public from possible exposure to pathogens. The bacteria that cause the closings are generally harmless, but they are present in large numbers in human and animal sewage. Their presence indicates the possible presence of disease-causing organisms.

The most common problem caused by swimming in contaminated water is gastroenteritis, which is contracted by swallowing water while swimming and can result in diarrhea, nausea, vomiting, and cramps. Even though gastroenteritis is generally not harmful to healthy adults, it can cause serious illness in children, the elderly, and people with autoimmune diseases, such as the human immunodeficiency virus and the acquired immunodeficiency syndrome.

The EPA established the Beaches Environmental Assessment and Coastal Health (BEACH) Program in 1997 to help reduce the risk of waterborne illness at the nation's beaches and recreational waters through improvements

that may cause illness. States use bacterial indicators to determine if oceans are safe for swimming or secondary contact recreation, such as waterskiing. Figure 1.5 in Chapter 1 shows the pathways of bacteria to surface waters. The most common sources of bacteria are urban runoff, inadequately treated human sewage, and runoff from pastures and feedlots (nonpoint sources), all of which were identified by several states as leading sources of ocean shoreline impairment. (See Figure 6.8.)

Turbidity, which is a measure of the relative clarity of water, is caused by suspended matter or other impurities that make the water look cloudy. These impurities may include clay, silt, finely divided organic and inorganic matter, plankton, and other microscopic organisms. It interferes with the transmission of light to underwater grasses and other plant life in need of this light. If the transmission of light is reduced because of heavy silt in the water, this can smother

in water protection programs and risk communication. Three years later, the BEACH Act of 2000 was signed into law. This law was an amendment to the Clean Water Act and required:

1. The EPA to issue new or revised water quality criteria for pathogens and pathogen indicators.

2. Coastal states to adopt these new or revised water quality standards.

3. The EPA to award grants to states and local governments to develop and implement beach monitoring and assessment programs.

The BEACH Act also required the EPA to prepare a progress report for Congress every four years. The first report was *Implementing the BEACH Act of 2000: Report to Congress* (October 2006. http://www.epa.gov/waterscience/beaches/report/full-rtc.pdf).

In *Implementing the BEACH Act*, the EPA determines that the major pollution sources responsible for beach closings and advisories in 2002 included runoff of storm water following rainfall (21%), sewage spills or overflows from various sources (13%), and unknown sources (43%). (See Figure 6.9.)

Table 6.5 shows the number of beaches surveyed for closings and advisories from 1997 through 2004. The number of beaches in the survey grew voluntarily from 1,021 in 1997 to 2,823 in 2002. Beginning in 2003, however, coastal states were required to report beach information to the EPA. Thus, the number surveyed in 2004 grew to 3,574.

Table 6.5 shows that from 1997 through 2004 approximately 26% of U.S. beaches were affected by advisories or closings. Nevertheless, the EPA notes in *Implementing the BEACH Act* that most of the advisories or closings lasted only one or two days. In 2004, for example, only 4% of possible open beach days were lost to advisories or closings.

As a result of the BEACH Act, the EPA has improved its Beach Advisory and Closing On-line Notification Web site (http://oaspub.epa.gov/beacon/beacon_national_page.main), which provides data to the public on beach advisories and closings. Along with collecting

more comprehensive data and strengthening water quality standards, the EPA is also working to improve pollution control efforts at the nation's beaches.

OCEAN POLLUTANTS—SOURCES AND EFFECTS

Any number of human-made materials or excessive amounts of naturally occurring substances can adversely affect marine and estuarine waters and their inhabitants. Because water is such an effective solvent and dispersant, it is difficult to track and quantify many pollutants known to have been discharged into marine and estuarine waters,

FIGURE 6.9

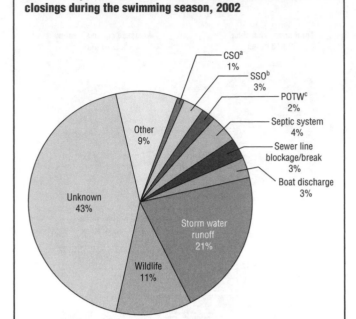

Sources of pollution that resulted in beach advisories or closings during the swimming season, 2002

[a]CSO is combined sewer overflow.
[b]SSO is sanitary sewer overflows.
[c]POTW is publicly owned treatment works; a wastewater treatment facility that is owned by a state or municipality.

SOURCE: "Figure 3.1. Sources of Pollution that Resulted in Beach Actions in 2002," in *Implementing the BEACH Act of 2000 Report to Congress*, U.S. Environmental Protection Agency, October 2006 http://www.epa.gov/waterscience/beaches/report/fullrtc.pdf (accessed February 6, 2007)

TABLE 6.5

Numbers and percentages of beaches affected by advisories or closings, 1997–2004

	Voluntary survey						Required reporting	
	1997	1998	1999	2000	2001	2002	2003	2004
Number of beaches	1,021	1,403	1,891	2,354	2,445	2,823	1,857*	3,574
Number of beaches affected by advisories or closings	230	353	459	633	672	709	395	942
Percentage of beaches affected by advisories or closings	23	25	24	27	27	25	21	26

*Incomplete data from 11 states; Environmental Protection Agency is working to complete data set.

SOURCE: "Table 3.2. National Health Protection Survey of Beaches Trends, 1997–2004," in *Implementing the BEACH Act of 2000 Report to Congress*, U.S. Environmental Protection Agency, October 2006, http://www.epa.gov/waterscience/beaches/report/full-rtc.pdf (accessed February 6, 2007)

and in many cases the source of pollution may be unknown. Some pollutants, such as oil spills, are easily detected the moment they enter the water. Others, such as toxic chemicals, are less obvious, and their presence may remain undetected until they cause extensive damage.

Oil Spills

Oil is one of the world's most important fuels. Its uneven distribution on the planet, however, forces its transport over the ocean, through pipelines, and over land. This inevitably results in accidents, some massive and some small, during drilling and transporting. In March 1967 the 118,285-ton supertanker *Torrey Canyon*, carrying oil from Kuwait, caused the world's first massive marine oil spill off the coast of England.

Oil spills are a dramatic form of water pollution—visible, immediate, and sometimes severe. The sight of dead and dying otters and birds covered with black film arouses instant sympathy, and the bigger the spill, the more newsworthy it is. Even though it is true that oil can have a devastating effect on marine life, the size of the spill itself is often not the determining factor in the amount of damage it causes. Other factors include the amount and type of marine life in the area and weather conditions that can disperse the oil. Despite the drama that tanker spills create, worldwide pollution from them is a relatively minor source of marine pollution. Tanker spills represent a small fraction of the oil released to the environment worldwide when compared with industry sources, nontanker shipping releases, and oil seepage from natural sources.

According to the EPA (March 9, 2006, http://www .epa.gov/oilspill/exxon.htm), when the supertanker *Exxon Valdez* ran into a reef in Prince William Sound, Alaska, in March 1989, more than eleven million gallons of oil spilled into one of the richest and most ecologically sensitive areas in North America. The *Exxon Valdez* Oil Spill Trustee Council (2007, http://www.evostc.state .ak.us/History/PWSmap.cfm) states that a slick (spill area) of approximately eleven thousand square miles—the size of Rhode Island and Maryland combined—threatened fish and wildlife. Otters died by the thousands, despite efforts by trained environmentalists and local volunteers to save them. Oil-soaked birds lined the shores, only to be eaten by larger predator birds, which then succumbed to dehydration and starvation because the ingested oil destroyed their metabolic systems.

OCEAN POLLUTION ACT OF 1990. In response to the *Exxon Valdez* disaster, Congress passed the Oil Pollution Act of 1990 (OPA). Most of the OPA provisions were targeted at reducing the number of spills and reducing the quantity of oil spilled. Among its provisions were the creation of a $1 billion cleanup-damage fund (the money comes from a tax on the petroleum industry), advance planning for controlling spills, stricter crew standards, and the requirement that new tankers have double hulls. When the exterior hull of a double-hulled tanker is punctured, the interior hull holding the oil may still remain intact. (The *Exxon Valdez* was not double-hulled.) The law requires older tankers to be fitted with double hulls by 2010. The OPA also:

- Compels the use of escort tugboats in certain harbors to assist tankers.

- Requires standards for tank levels and pressure-monitoring devices to detect leaks in cargo tanks.

- Requires the U.S. Coast Guard to establish minimum standards for overfill devices to prevent overfill oil spills. (An overfill oil spill is the result of too much oil being pumped into a tanker during a transfer from a facility to a tanker or between two tankers. On occasion, overfill spills have involved large quantities of oil.)

DECLINES IN OIL SPILLS. Declines in oil spills are being seen on a global scale. The International Tanker Owners Pollution Federation reports that between 1970 and 1979 the average incident rate for large spills (over 700 metric tons or approximately 215,000 gallons) from the worldwide tanker industry was 25.2 spills per year. (See Figure 6.10.) Between 1980 and 1989 the average rate dropped to 9.3 spills per year. From 1990 to 1999 the global spill rate declined further to 7.8 spills per year and from 2000 through 2006 to 3.7 spills per year. Larger spills, such as these, are most often caused by collisions and groundings. Smaller spills most often occur because of routine operations, such as loading in ports or at oil terminals.

The decline in oil spills in U.S. waters is even more dramatic. Figure 6.11 shows the number of oil spills over one thousand gallons in U.S. waters from 1973 to 2004. Most spills are not shown on this graph because they are smaller than one thousand gallons. According to the report *U.S. Coast Guard Polluting Incident Compendium: Cumulative Data and Graphics for Oil Spills 1973–2004* (2006, http://www .uscg.mil/hq/gm/nmc/response/stats/Summary.htm), 88% of all spills from 1973 to 2004 were between one and one hundred gallons. Spills over one hundred thousand gallons have not occurred in U.S. waters since 1996, and before that, since 1990.

Although oil tanker spills are highly visible cases of pollution entering the ocean, the U.S. Department of the Interior's Minerals Management Service reports in *OCS Oil Spill Facts* (September 2002, http://www.mms.gov/ stats/PDFs/2002OilSpillFacts.pdf) that the largest input of oil into marine environments is natural seepage (naturally occurring oil in the ground that moves through the soil and into the water). In North America natural seepage contributes 63% of the total marine oil input. Twenty-two

FIGURE 6.10

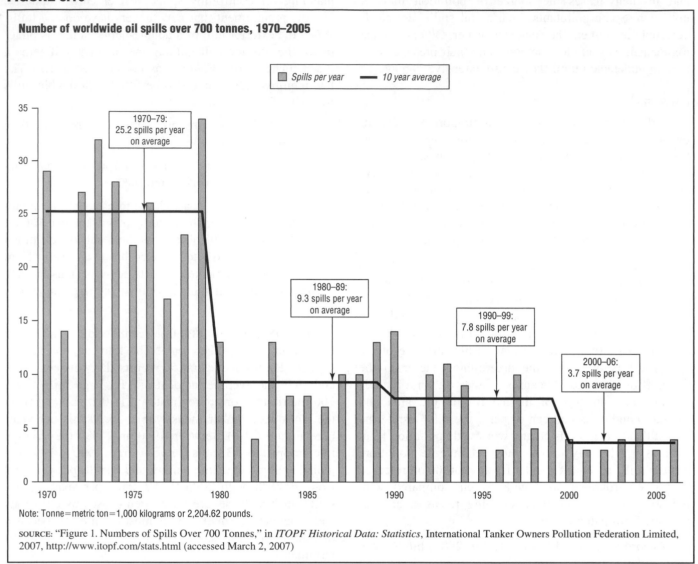

Number of worldwide oil spills over 700 tonnes, 1970–2005

Note: Tonne=metric ton=1,000 kilograms or 2,204.62 pounds.

SOURCE: "Figure 1. Numbers of Spills Over 700 Tonnes," in *ITOPF Historical Data: Statistics*, International Tanker Owners Pollution Federation Limited, 2007, http://www.itopf.com/stats.html (accessed March 2, 2007)

percent of the oil found in the marine waters off the coast of North America is due to municipal and industrial waste and runoff. Marine transportation is responsible for only 3%, although worldwide, it is responsible for 33%.

Marine Debris

NOAA (March 27, 2007, http://marinedebris.noaa .gov/whatis/welcome.html) defines marine debris as "any man-made object discarded, disposed of, or abandoned that enters the coastal or marine environment. It may enter directly from a ship, or indirectly when washed out to sea via rivers, streams and storm drains." The effects of marine debris can be both costly to coastal communities and dangerous to humans and aquatic life. Certain types of marine debris, such as broken glass and medical waste, can pose a serious threat to public health, causing beach closures and swimming advisories and robbing coastal communities of significant tourism dollars.

In December 2006 the Marine Debris Research, Prevention, and Reduction Act became law. The purpose of the act was to create a Marine Debris Research, Prevention, and Reduction Program within NOAA and the Coast Guard to help identify, determine sources of, assess, reduce, and prevent marine debris and its adverse impacts on the marine environment and navigation safety.

CRUISE SHIP WASTE. Laurie Asseo reports in "Cruise Line Fined Millions for Dumping" (*Milwaukee Journal Sentinel*, July 22, 1999) that in 1999 the Royal Caribbean, one of the world's largest cruise lines, pleaded guilty in federal court to dumping oil and hazardous chemicals in U.S. waters and lying about it to the Coast Guard. It agreed to pay a record $18 million fine for polluting waters. This was besides the $9 million in criminal fines the company agreed to pay in a previous plea agreement. Asseo notes that six other cruise lines have pleaded guilty to illegal waste dumping since 1993 and have paid fines of up to $1 million. These cases focus attention on the difficulties of regulating the expanding cruise line industry, because most major ships sailing out of U.S. ports are registered in foreign countries.

FIGURE 6.11

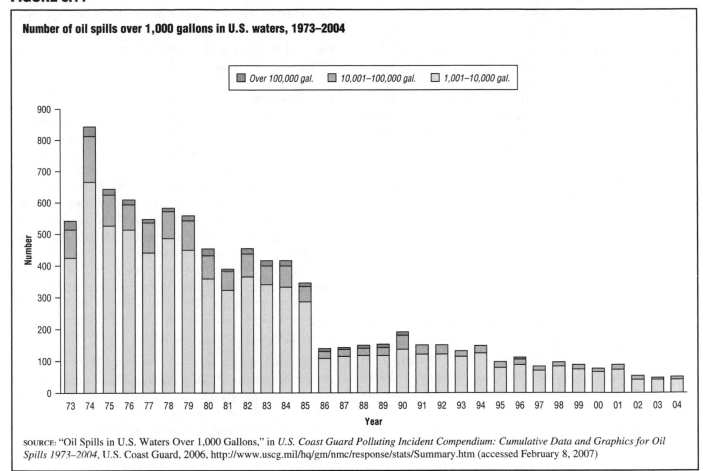

Number of oil spills over 1,000 gallons in U.S. waters, 1973–2004

Legend: ■ Over 100,000 gal. ■ 10,001–100,000 gal. □ 1,001–10,000 gal.

SOURCE: "Oil Spills in U.S. Waters Over 1,000 Gallons," in *U.S. Coast Guard Polluting Incident Compendium: Cumulative Data and Graphics for Oil Spills 1973–2004*, U.S. Coast Guard, 2006, http://www.uscg.mil/hq/gm/nmc/response/stats/Summary.htm (accessed February 8, 2007)

In "Cruise Ship Water Discharges" (March 23, 2007, http://www.epa.gov/owow/oceans/cruise_ships/), the EPA reports that there are more than 230 cruise ships operating worldwide. Feeding and housing thousands of people on each of these vessels means that a great deal of waste is generated while at sea. The EPA notes that "some of the waste streams generated by cruise ships include bilge water (water that collects in the lowest part of the ship's hull and may contain oil, grease, and other contaminants), sewage, graywater (waste water from showers, sinks, laundries and kitchens), ballast water (water taken onboard or discharged from a vessel to maintain its stability), and solid waste (food waste and garbage)."

The EPA is conducting a variety of activities to address the problem of cruise ship waste. Among these activities the agency is assessing the need for additional standards for waste discharge from cruise ships operating in Alaska and is preparing a cruise ship waste assessment report (September 14, 2006, http://www.epa.gov/owow/oceans/cruise_ships/disch_assess.html). The assessment report will characterize cruise ship waste, analyze existing programs for managing those wastes, and determine whether better management of those wastes is needed.

PLASTICS. Plastics such as bags, containers, bottle caps, and beverage carriers are dumped daily from ocean-going vessels, commercial and recreational fishing boats, offshore oil and gas platforms, and military ships. Other types of plastic debris—factory wastes, sewer overflows, illegal garbage dumping, and human littering—come from land sources. Thousands of seabirds and marine animals die each year as a result of ingesting or becoming entangled in this plastic.

Another concern is commercial fishing nets. Once made of natural materials, these nets are now made mainly of durable, nondegradable plastic. When they are lost or discarded in the ocean, they pose a floating hazard to seals, dolphins, whales, and diving birds, which can become entangled in the nets. In 1988 thirty-one nations ratified an agreement making it illegal for their ships to dump plastic debris, including fishing nets, into the ocean. As part of that agreement, the United States enacted the Marine Plastics Pollution Research and Control Act, which went into effect in 1989. Among other regulations, the act imposes a $25,000 fine for each violation.

Plastic pellets are the raw materials that are melted and molded to create plastic products. According to Paul

Watson, in "The Plastic Sea" (July 24, 2006, http://www.seashepherd.org/editorials/editorial_060724_1.html), sixty billion pounds of resin pellets are manufactured in the United States annually. The two primary ways that these pellets enter the ocean are direct spills during cargo handling operations at ports or spills at sea, and storm water discharges that carry the pellets from industrial sites. Plastic pellets may persist in the water environment for years, depending on the resin type, the amount and types of pellet additives, and how the pellets react to sunlight, wave action, and weathering. Although pellets have been found in the stomachs of wildlife, primarily seabirds and sea turtles, their effects have not been clearly demonstrated to be harmful.

Since 1991 the Society of Plastics Industries (SPI), the major national trade association for manufacturers who make plastic products in the United States, has been working with the EPA to identify and minimize the sources of plastic pellet entry into the ocean. In July 1991 the SPI instituted Operation Clean Sweep (http://www.opcleansweep.org/), an industry-wide education campaign to encourage members to adopt the SPI 1991 Pellet Retention Environmental Code and the 1992 Processor's Pledge aimed at committing the U.S. plastics industry to total pellet containment.

GHOST FISHING. Another important problem is ghost fishing. This is the entrapment of fish and marine mammals by lost or abandoned nets, pots, fishing line, bottles, and other discarded objects. When marine creatures are entangled in old six-pack beverage binders or caught in abandoned fishing nets, they suffer and may die.

Ocean Dumping

In 1972 Congress enacted the Marine Protection, Research, and Sanctuaries Act (also known as the Ocean Dumping Act) to prohibit the dumping into the ocean of material that will unreasonably degrade or endanger human health or the marine environment. The act applies to waters within two hundred miles of the U.S. coast and was amended in 1988 to prohibit dumping industrial waste and sewage sludge into the ocean. As a result, the only ocean dumping allowed as of early 2007 was dredged material from the bottom of water bodies to maintain navigation channels and berthing areas. To dump dredged material in the ocean, a permit must be obtained from the EPA.

ALGAL BLOOMS

Algae are plantlike organisms that manufacture their own food via photosynthesis. Most algal species in U.S. coastal waters are not harmful and serve as the energy producers at the base of the food chain. However, sometimes algae may grow fast and bloom, creating dense, ~~[patch]~~ patches near the water surface. Red tide is a common name for events in which certain algae containing reddish pigments bloom so that the water appears to be red. Often, these particular species are toxic to humans and wildlife, but not all species of algae are toxic, nor do all species impart color to the water during blooms.

Eutrophication and Hypoxia

Algal blooms often occur because an abundance of plant nutrients, such as nitrates and phosphates, have entered the water. Thick layers of algae block the sunlight from reaching the algae and other plant life below, so those organisms die. (See Figure 6.12.) When the nutrients run out, much of the rest of the algae die. Bacteria and other decomposers feed on the dead algae, using oxygen in the water as they break down the tissues. This process, in which an overenrichment of a water body with nutrients results in an excessive growth of organisms and a resultant depletion of oxygen concentration, is called eutrophication. The deficiency of oxygen in the water is called hypoxia, and it is a condition that can have severe effects on local ecosystems.

Hypoxia kills most of the sessile (permanently attached) bottom-dwelling benthic organisms in a body of water, such as oysters and clams; aquatic animals that swim or crawl, such as fish, shrimp, and crabs, either leave the area or die as shown in Figure 6.12. For this reason, areas where hypoxic conditions exist are frequently referred to as dead zones. Hypoxia is a worldwide problem that often occurs where rivers carrying large amounts of agricultural runoff empty into lakes, estuaries, and the ocean.

In 1998 Congress passed the Harmful Algal Bloom and Hypoxia Research Act. The act requires the formation of a federal multiagency task force to investigate the problem and report back to Congress with a plan and recommendations to address harmful algal blooms and hypoxia. The Harmful Algal Bloom and Hypoxia Amendments Act of 2004 reauthorized the 1998 act.

EUTROPHICATION AND HYPOXIA IN THE GULF OF MEXICO. One location in the United States where hypoxia occurs is the Gulf of Mexico, off the Louisiana coast. According to the USGS in the fact sheet "Restoring Life to the Dead Zone: Addressing Gulf Hypoxia, a National Problem" (June 2000, http://www.nwrc.usgs.gov/factshts/016-00/016-00.htm), the Gulf's hypoxic zone is comparable to the largest hypoxic areas in the world, such as those in the Black and Baltic seas. The Gulf of Mexico hypoxic zone is approximately six to seven thousand square miles of water where the oxygen level is below two ppm. Under normal conditions dissolved oxygen levels would be five to six ppm.

The zone is caused by harmful algal blooms that are believed to be the result of the discharge of nutrients from the Mississippi River watershed into the Gulf of

FIGURE 6.12

Eutrophication and hypoxia in a body of water

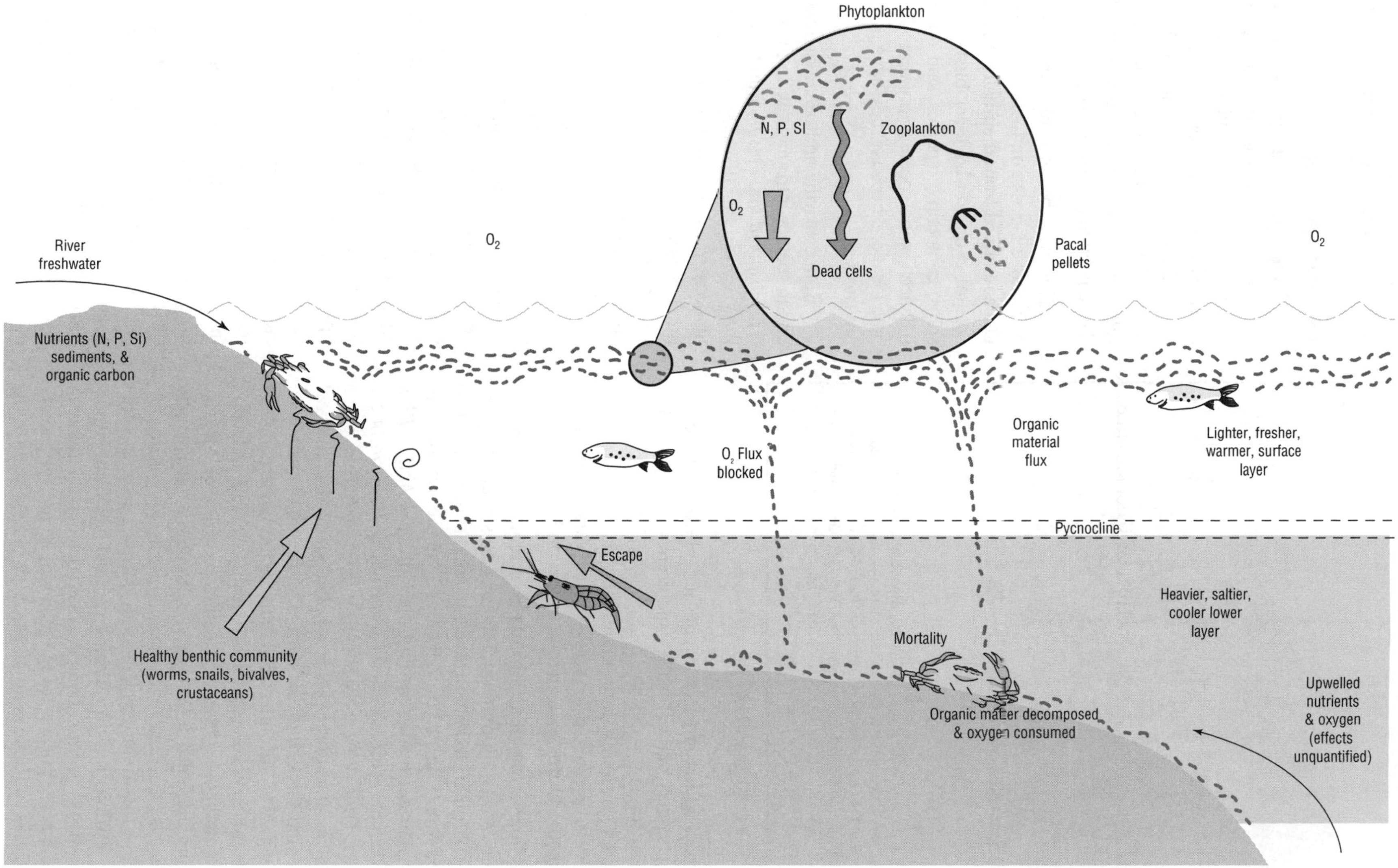

River
freshwater

Nutrients (N, P, Si)
sediments, &
organic carbon

Phytoplankton

Zooplankton

N, P, Si

O₂

Dead cells

Pacal
pellets

O₂

Lighter, fresher,
warmer, surface
layer

Organic
material
flux

Pycnocline

Heavier, saltier,
cooler lower
layer

Upwelled
nutrients
& oxygen
(effects
unquantified)

Organic matter decomposed
& oxygen consumed

Mortality

O₂ Flux
blocked

Escape

Healthy benthic community
(worms, snails, bivalves,
crustaceans)

O₂

SOURCE: *Mississippi River Basin & Gulf of Mexico Hypoxia: What is Hypoxia?* U.S. Environmental Protection Agency, 2005, http://www.epa.gov/msbasin/taskforce/euro.htm (accessed January 10, 2007)

FIGURE 6.13

Interior watersheds of the Mississippi River Basin

SOURCE: "Interior Watersheds of the Mississippi River Basin, the Source of Materials Causing the 6,000- to 7,000-Square-Mile 'Dead Zone,' or Hypoxia in the Gulf of Mexico," in *Restoring Life to the Dead Zone: Addressing Gulf Hypoxia, a National Problem*, U.S. Department of the Interior, U.S. Geological Survey, June 2000, http://www.nwrc.usgs.gov/factshts/016-00.pdf (accessed January 4, 2007)

Mexico. The nutrients (nitrogen and phosphorus) come from fertilizers, animal waste, and domestic sewage. The nitrate-nitrogen level in the main stem of the Mississippi River, which drains thirty-one states, has doubled since the 1950s. Figure 6.13 shows the Mississippi Basin watershed and the states whose rivers drain into it.

To correct the situation and as a requirement of the Harmful Algal Bloom and Hypoxia Research Act, the EPA, six other federal agencies, nine states, and two Native American tribes developed an action plan to reduce nutrient loads reaching the Gulf: the *Action Plan for Reducing, Mitigating, and Controlling Hypoxia in the Northern Gulf of Mexico* (January 2001, http://www.epa.gov/msbasin/taskforce/pdf/actionplan.pdf). It has the goal of reducing the size of the hypoxic zone by over 68% no later than 2015. The plan also calls for implementation of nutrient management strategies to achieve a 30% reduction in the amount of nutrients reaching the Gulf of Mexico. Reducing nutrients in the water, particularly nitrogen and phosphorus, will help reverse the hypoxia. This is accomplished primarily by implementing farming practices that reduce fertilizer runoff, restoring wetland areas, and restoring riverbanks. Information generated through the research and monitoring portions of the plan will be used to modify future goals and actions as necessary. As of early 2007, the

Basin Expert Review report was expected to be completed in the spring or summer of 2007, with a revised action plan due in 2008 (January 22, 2006, http://www.epa.gov/msbasin/taskforce/pdf/timeline_process01_06.pdf).

EXOTIC SPECIES

Exotic species are plants, animals, and microbes that have been carried from one geographic region to another, either intentionally or unintentionally. Unintentional introduction includes transport in ballast water of ships or as pests on imported fruits, vegetables, and animals or animal products. Before modern times, movement from one geographical region to another was infrequent and slow, allowing time for the ecology to absorb and counterbalance the newcomers.

However, because of rapid transport, organisms can now move across continents in a matter of hours or days. Once removed from their natural ecological system, where eons of evolution have established predator-prey relationships, competitive species, and other devices to maintain balance, exotic species may reproduce unchecked in their new locations because they have no natural competitors or predators.

Both estuarine and ocean habitats have suffered from exotic species introduction. In the Chesapeake Bay, MSX (*Haplosporidium nelsoni*) and Dermo (*Perkinsus marinus*), two oyster diseases that have ravaged oyster populations, came to the Bay with oysters introduced from other regions. The coral reefs in the Northern Mariana Islands are being decimated by the introduction of the crown-of-thorns starfish. The green crab introduced from the Baltic Sea to the shores of New England occurs in such high numbers that the crab is believed to be eating young scallops and other valuable seafood.

Passage of the Nonindigenous Aquatic Nuisance Prevention and Control Act of 1990 was a first step in attempting to prevent species migration. This legislation authorized the Fish and Wildlife Service and NOAA to adopt regulations to prevent the unintentional introduction of aquatic nuisance species. In 1999 the Invasive Species Council was created by presidential executive order to oversee efforts to control unwanted exotic species. The council is chaired jointly by the secretaries of interior, agriculture, and commerce. Council members include the secretaries of state, treasury, and transportation, and the administrator of the EPA. To date, a variety of laws have been enacted and legislation has been introduced to help research this problem and find ways to abate and control it. The Northeast Midwest Institute, in "Biological Pollution" (February 20, 2007, http://www.nemw.org/biopollute.htm), provides a listing of the proposed and enacted legislation.

CHAPTER 7
WETLANDS

WHAT ARE WETLANDS?

Wetlands are transition zones between land and aquatic systems where the water table is usually near or at the surface, or the land is covered by shallow water. Wetlands range in size from less than one acre to thousands of acres and can take many forms, some of which are immediately recognizable as "wet." Other wetlands appear more like dry land, and are wet during only certain seasons of the year, or at several year intervals. The U.S. Army Corps of Engineers reports that most U.S. wetlands lack surface water and waterlogged soils during at least part of each growing season.

Some of the more commonly recognized types of wetlands are marshes, bogs, and swamps. Marshes are low-lying wetlands with grassy vegetation. Bogs are wetlands that accumulate wet, spongy, acidic, dead plant material called peat. Shrubs, mosses, and stunted trees may also grow in bogs. Swamps are low-lying wetlands that are seasonally flooded; they have more woody plants than marshes and better drainage than bogs.

According to Thomas E. Dahl, in *Status and Trends of Wetlands in the Conterminous United States 1998 to 2004* (2006, http://wetlandsfws.er.usgs.gov/status_trends/national_reports/trends_2005_report.pdf), there were an estimated 107.7 million acres of wetlands in the forty-eight coterminous states in 2004. This constituted about 5.5% of the total land area, whereas deepwater (rivers and lakes) constituted 1% and upland ("dry" land) constituted most of the total land area at 93.5%. (See Figure 7.1.) Of the wetland areas, 95% were freshwater and 5% were estuarine (coastal saltwater). (See Figure 7.2.) Dahl does not include data on Alaska and Hawaii, but the U.S. Environmental Protection Agency (EPA) notes in "Wetlands: Status and Trends" (February 22, 2006, http://www.epa.gov/OWOW/wetlands/vital/status.html) that in the 1980s an estimated 170 to 200 million acres of wetlands existed in Alaska—covering slightly more than half of the state—and Hawaii had 52,000 acres.

Hydrology and Wetland Formation

Wetlands are distributed unevenly, but occur in every state and U.S. territory. (See Table 7.1.) They are found wherever climate and landscape cause groundwater to discharge to the land surface or prevent rapid drainage from the land surface so that soils are saturated for some time.

In wetlands, when the soil is flooded or saturated, the oxygen used by the microbes and other decomposers in the water is slowly replaced by oxygen in the air, because oxygen moves through water about ten thousand times slower than through air. Thus, all wetlands have one common trait: hydric (oxygen-poor) soils. As a result, plants that live in wetlands have genetic adaptations in which they are able to survive temporarily without oxygen in their roots, or they are able to transfer oxygen from the leaves or stem to the roots. This anaerobic (without oxygen) condition causes wetland soils to have the sulfurous odor of rotten eggs.

Local hydrology (the pattern of water flow through an area) is the primary determinant of wetlands. Wetlands can receive groundwater in-flow, recharge groundwater, or experience both inflow and outflow at different locations. Figure 7.3 illustrates water movement in several different wetland situations. Part A in Figure 7.3 shows how wetlands do not always occupy low points and depressions in the landscape. They can occur in flat areas that have complex underground water flow.

Part B in Figure 7.3 shows a fen, which is a type of wetland that accumulates peat deposits like bogs do. Fens, however, are less acidic than bogs and receive most of their water from groundwater rich in calcium and magnesium. As part B shows, fens occur on slopes at groundwater seepage faces and are subject to a continuous supply of the chemicals that are dissolved in the groundwater.

FIGURE 7.1

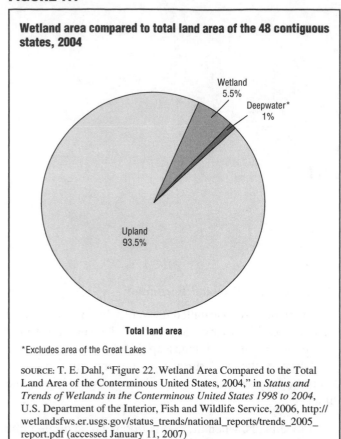

Wetland area compared to total land area of the 48 contiguous states, 2004

Wetland
5.5%

Deepwater*
1%

Upland
93.5%

Total land area

*Excludes area of the Great Lakes

SOURCE: T. E. Dahl, "Figure 22. Wetland Area Compared to the Total Land Area of the Conterminous United States, 2004," in *Status and Trends of Wetlands in the Conterminous United States 1998 to 2004*, U.S. Department of the Interior, Fish and Wildlife Service, 2006, http://wetlandsfws.er.usgs.gov/status_trends/national_reports/trends_2005_report.pdf (accessed January 11, 2007)

FIGURE 7.2

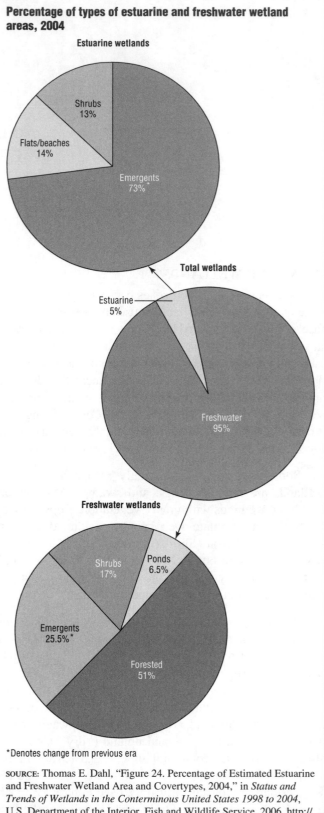

Percentage of types of estuarine and freshwater wetland areas, 2004

Estuarine wetlands

Shrubs
13%

Flats/beaches
14%

Emergents
73%

Total wetlands

Estuarine
5%

Freshwater
95%

Freshwater wetlands

Shrubs
17%

Ponds
6.5%

Emergents
25.5%*

Forested
51%

*Denotes change from previous era

SOURCE: Thomas E. Dahl, "Figure 24. Percentage of Estimated Estuarine and Freshwater Wetland Area and Covertypes, 2004," in *Status and Trends of Wetlands in the Conterminous United States 1998 to 2004*, U.S. Department of the Interior, Fish and Wildlife Service, 2006, http://wetlandsfws.er.usgs.gov/status_trends/national_reports/trends_2005_report.pdf (accessed January 11, 2007)

Land along the sides of streams or rivers receives a continuous water supply and is ideal for wetland growth. It may also receive some groundwater discharge, as shown in part C in Figure 7.3. Bogs, shown in part D in Figure 7.3, are wetlands normally found on uplands or extensive flatlands. Most of their water and chemistry comes from precipitation.

Riverine (areas along streams, rivers, and irrigation canals) and coastal area wetlands are highly subject to periodic water level changes. Coastal area wetlands, for example, are affected by predictable tidal cycles. Other coastal and riverine wetlands are highly dependent on flooding and seasonal water level changes. Some examples are the floodplains of the Illinois and Missouri rivers.

TYPES OF WETLANDS

A wide variety of wetlands exist across the United States because of regional and local differences in hydrology, water chemistry, vegetation, soils, topography, and other factors. There are two large groups of wetlands: estuarine (coastal) and freshwater (inland). (See Figure 7.2.) Estuarine wetlands are linked to estuaries and oceans and are places where fresh- and saltwater mix, such as a bay or where a river enters the ocean. In estuaries the environment is one of ever-changing salinity and temperature. The water level fluctuates in response to

wind and tide. Examples of estuarine wetlands are saltwater marshes and mangrove swamps.

TABLE 7.1

Location of various wetland types

Wetland type	Primary regions	States
Inland freshwater marsh	Dakota-Minnesota drift and lake bed; Upper Midwest; and Gulf Coastal Flats	North Dakota, South Dakota, Nebraska, Minnesota, Florida
Inland saline marshes	Intermontane; Pacific Mountains	Oregon, Nevada, Utah, California
Bogs	Upper Midwest; Gulf-Atlantic Rolling Plain; Gulf Coastal Flat; Atlantic Coastal Flats	Wisconsin, Minnesota, Michigan, Maine, Florida, North Carolina
Tundra	Central Highland and Basin; Arctic Lowland; and Pacific Mountains	Alaska
Shrub swamps	Upper Midwest; Gulf Coastal Flats	Minnesota, Wisconsin, Michigan, Florida, Georgia, South Carolina, North Carolina, Louisiana
Wooded swamps	Upper Midwest; Gulf Coastal Flats; Atlantic Coastal Flats; and Lower Mississippi Alluvial Plain	Minnesota, Wisconsin, Michigan, Florida, Georgia, South Georgia, South Carolina, North Carolina, Louisiana
Bottom land hardwood	Lower Mississippi Alluvial Plain; Atlantic Coastal Flats; Gulf-Atlantic Rolling Plain; and Gulf Coastal Flats	Louisiana, Mississippi, Arkansas, Missouri, Tennessee, Alabama, Florida, Georgia, South Carolina, North Carolina, Texas
Coastal salt marshes	Atlantic Coastal Zone; Gulf Coastal Zone; Eastern Highlands; Pacific Moutains	All Coastal states, but particularly the Mid- and South Atlantic and Gulf Coast states
Mangrove swamps	Gulf Coastal Zone	Florida and Louisiana
Tidal freshwater wetlands	Atlantic Coastal Zone and Flats; Gulf Coastal Zone and Flats	Louisiana, Texas, North Carolina, Virginia, Maryland, Delaware, New Jersey, Georgia, South Carolina

SOURCE: "Table 3, Locations of Various Wetland Types in the United States," in *Wetlands: Their Use and Regulation*, U.S. Congress, Office of Technology Assessment, March 1984, http://govinfo.library.unt.edu/ota/Ota_4/DATA/1984/8433.PDF (accessed January 11, 2007)

FIGURE 7.3

Examples of water sources for wetlands

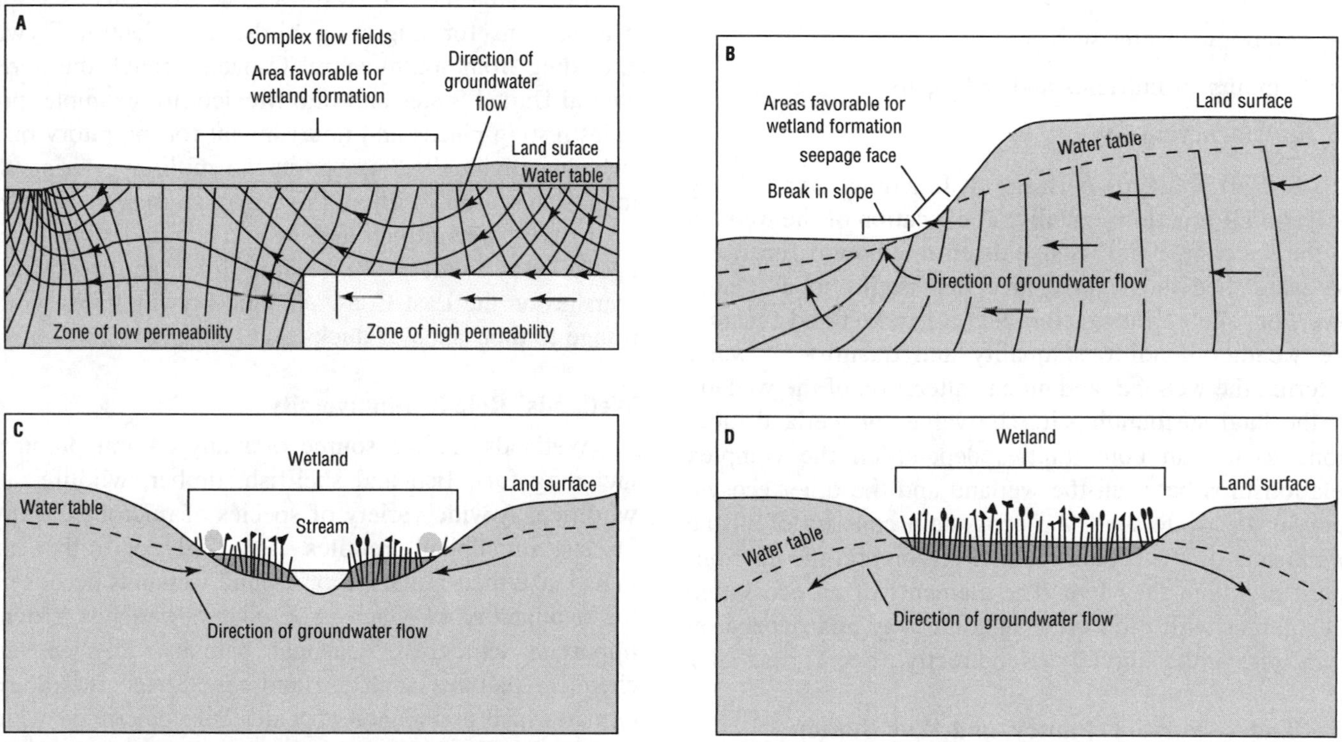

The source of water to wetlands can be from groundwater discharge where the land surface is underlain by complex groundwater flow fields (A), from groundwater discharge at seepage faces and at breaks in slope of the water table (B), from streams (C), and from precipitation in cases where wetlands have no stream inflow and groundwater gradients slope away from the wetland (D).

SOURCE: Thomas C. Winter et al., "Figure 17," in *Ground Water and Surface Water: A Single Resource*, U.S. Department of the Interior, U.S. Geological Survey, 1998, http://pubs.usgs.gov/circ/circ1139/pdf/circ1139.pdf (accessed January 4, 2007)

The most common location of freshwater wetlands is the floodplains of rivers and streams, the margins of lakes and ponds, and isolated depressions surrounded by dry land. Some examples of inland wetlands are the Florida Everglades, wet meadows, swamps, fens, bogs, prairie potholes, playa lakes, and wet tundra.

Wetlands are further divided by their vegetation. Emergent wetlands (marshes and wet meadows) are dominated by grasses, sedges, and other herbaceous (non-woody) plants. Emergent wetlands account for 73% of estuarine wetlands, even though they represent only 25.5% of freshwater wetlands. (See Figure 7.2.) Shrub wetlands (including shrub swamps and bogs), which are characterized by low-to-medium-height woody plants, make up 13% of estuarine wetlands and account for 17% of freshwater wetlands. Forested wetlands, mostly wooded swamps and bottomland hardwood forests, are dominated by trees and account for 51% of freshwater wetlands. (Bottomland hardwood forests are generally found along the edges of lakes and rivers and in sinkholes.)

HOW WETLANDS FUNCTION

Wetlands provide essential ecological functions that benefit people and the ecological systems surrounding the wetlands, as well as the wetland itself. The plants, microbes, and animals in wetlands are all key players in the water, nitrogen, carbon, and sulfur cycles.

Wetland functions fit into several broad categories:

- High plant productivity
- Temporary water storage
- Trapping of nutrients and sediments
- Soil anchoring

Not all wetlands perform all functions, nor do they perform all functions equally. The location of the wetland in the watershed and its size determine how it functions. (A watershed is the land area that drains to a stream, river, or lake.) Other factors that affect wetland function are weather conditions, quality and quantity of water entering the wetland, and human alteration of the wetland or the land surrounding it. The values of wetland functions to human communities depend on the complex relationships between the wetland and the other ecosystems in the watershed. An ecosystem consists of all the organisms in a particular area or region and the environment in which they live. The elements of an ecosystem all interact with each other in some way and depend on each other either directly or indirectly. (See Figure 7.4.)

Wetlands—Nursery, Pantry, and Way Station

Wetlands are diverse and rich ecosystems, which provide food and shelter to many different plants and animals. The combination of shallow water, high nutrient levels, and primary productivity (plant growth and reproduction) is perfect for the development of organisms that form the base of the food chain. The water, dense plants, their root mats, and decaying vegetation are food and shelter for the eggs, larvae, and juveniles of many species. Smaller animals avoid predators by hiding among the vegetation while they wait to prey on still smaller organisms. Fish of all sizes seek the warmer, shallow waters to mate and spawn, leaving their young to grow on the rich diet provided by the wetlands. Food and organic material that is flushed out of wetlands and into streams and rivers during periods of high water flow feed downstream aquatic systems, including commercial and sport fisheries.

Estuarine marshes, for example, are among the most productive natural ecosystems in the world. They produce huge amounts of plant leaves and stems that make up the base of the food chain. When the plants die, decomposers such as bacteria in the water break them down to detritus (small particles of organic material). Algae that grow on plants and detritus are the principal foods for shellfish such as oysters and clams, crustaceans such as crabs and shrimp, and small fish. Small fish are the food for larger commercial species such as striped bass and bluefish. (See Figure 7.5.) The EPA states in *Wetlands Functions and Values* (March 19, 2007, http://www.epa.gov/watertrain/wetlands/module08.htm) that "the fish and shellfish that depend on wetlands for food or habitat constitute more than 75% of the commercial and 90% of the recreational harvest."

Both estuarine and freshwater wetlands also serve as way stations for migrating birds. The Central Flyway extending from south-central Canada through the north-central United States and into Mexico, for example, provides resting places and nourishment for migratory birds (which individually number in the millions) during the migration season. Without this stopover area, the flight to their Arctic breeding grounds would be impossible. Chesapeake Bay with its extensive tidal and freshwater marshes on the East Coast Atlantic Flyway gives winter refuge to thousands of ducks and geese.

Wetlands' Role in Biodiversity

Wetlands are the source of many natural products, including furs, fish and shellfish, timber, wildlife, and wild rice. A wide variety of species of microbes, plants, insects, amphibians, reptiles, fish, birds, and other animals make their homes in or around wetlands because of the availability of water. For others, wetlands provide important temporary seasonal habitats. Physical and chemical features such as landscape shape (topology), climates, and abundance of water help determine which species live in which wetland.

In "Wetlands and People" (February 22, 2006, http://www.epa.gov/owow/wetlands/vital/people.html), the EPA notes that "more than one-third of the United States' threatened and endangered species live only in wetlands." When wetlands are removed from a watershed or are damaged by human activity, the biological health of the watershed declines. Wetland health has a commercial impact as well. Dahl indicates that 75% of the fish and shellfish commercially

FIGURE 7.4

Ecological value of wetland processes

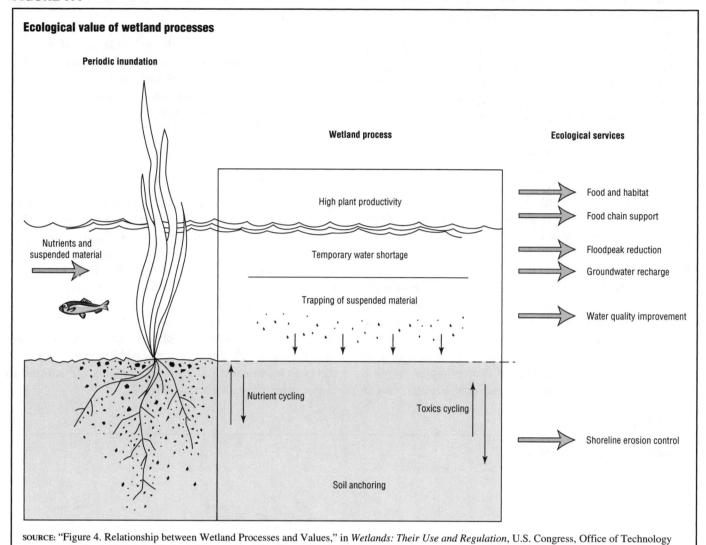

SOURCE: "Figure 4. Relationship between Wetland Processes and Values," in *Wetlands: Their Use and Regulation*, U.S. Congress, Office of Technology Assessment, March 1984, http://govinfo.library.unt.edu/ota/Ota_4/DATA/1984/8433.PDF (accessed January 11, 2007)

FIGURE 7.5

Coastal wetlands produce food for fish and shellfish

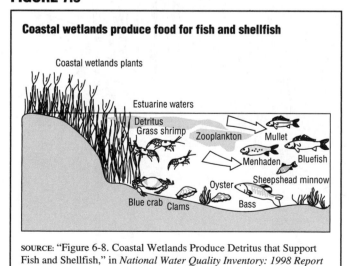

SOURCE: "Figure 6-8. Coastal Wetlands Produce Detritus that Support Fish and Shellfish," in *National Water Quality Inventory: 1998 Report to Congress*, U.S. Environmental Protection Agency, Office of Water, June 2000, http://www.epa.gov/305b/98report/chap6.pdf (accessed January 4, 2007)

harvested in the United States and up to 90% of the recreational fish rely directly or indirectly on wetlands for their survival. Dahl also notes that in 2004, 72% of freshwater mussels were imperiled and 39% of native freshwater fish species were at risk of extinction.

Waterfowl are birds such as ducks and geese that spend much of their lives in wetlands, lakes, rivers, and streams. The well-being of waterfowl populations is tied directly to the status and abundance of wetland habitats. According to the U.S. Fish and Wildlife Service's (USFWS) Division of Bird Habitat Conservation (December 12, 2006, http://www.fws.gov/birdhabitat/NAWMP/index.shtm), waterfowl are the most well-known and economically important group of migratory birds in North America. By 1985 (when waterfowl populations had decreased to record lows) approximately 3.2 million people were spending nearly $1 billion annually to hunt waterfowl. In addition, about 18.6 million people spent $2 billion on "waterfowl-watching" activities, such as observing and photographing them.

Measures to preserve and protect the waterfowl population include the North American Waterfowl Management Plan. A joint strategy adopted by the governments of the United States, Canada, and Mexico, the plan established an international committee with six representatives from each of the three countries. Its purpose is to provide a forum for discussion of major, long-term international waterfowl issues and to make recommendations to the directors of the three countries' national wildlife agencies. The Division of Bird Habitat Conservation notes that as of the end of 2006, $4.5 billion had been invested under the plan to protect, restore, and/or enhance 15.7 million acres of waterfowl habitat. North American Waterfowl Management Plan projects also target all wetland-associated species in their conservation efforts.

Water Storage

Wetlands absorb water, much like sponges. By temporarily storing runoff and flood waters, wetlands help protect adjacent and downstream property owners from flood damage. Wetland plants slow the flow of water, which contributes to the wetland's ability to store it. The combined effects of storing and slowing the flow of water permit it to percolate through the soil into groundwater, which recharges aquifers, and to move through the watershed with less speed and force.

Wetlands are particularly valuable in urban areas because paved and other impermeable surfaces shed water, increasing the rate, velocity, and volume of runoff so that the risk of flood damage increases. Loss or degradation of wetlands indirectly intensifies flooding by eliminating absorption of the peak flows and gradual release of floodwaters.

Nutrient and Sediment Control

Figure 7.6 shows how wetlands improve the quality of water. Wetlands act like natural water filters. When water is stored or slowed down in a wetland by the plants and root masses that grow there, sediment settles out and remains in the wetland so that the water leaving the area is much less cloudy than the water that entered. The loss of cloudiness or turbidity has important consequences for both human health and the ecological health of the watershed. Turbidity has been implicated in disease outbreaks in drinking water. Furthermore, turbid water bearing silt has been responsible for smothering plants and animals in rivers, streams, estuaries, and lakes.

FIGURE 7.6

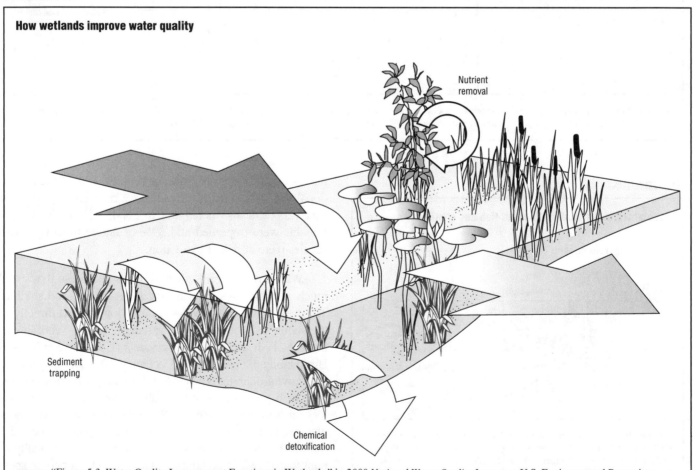

How wetlands improve water quality

Nutrient removal

Sediment trapping

Chemical detoxification

SOURCE: "Figure 5-3. Water Quality Improvement Functions in Wetlands," in *2000 National Water Quality Inventory*, U.S. Environmental Protection Agency, Office of Water, August 2002, http://www.epa.gov/305b/2000report/chp5.pdf (accessed January 5, 2007)

Wetlands can also trap nutrients (phosphorous and nitrogen) that are dissolved in the water or attached to the sediment. Nutrients are either stored in the wetland soil or used by the plants to enhance growth. If too much nutrient material reaches rivers, streams, lakes, and reservoirs, it can cause eutrophication, resulting ultimately in the death of many aquatic organisms. (See Figure 6.12 in Chapter 6 and the related discussion.)

Soil Anchoring

Wetlands also play an important role in soil anchoring. The thick mesh of wetland vegetation and roots acts like a net and helps hold soil in place even during periods of relatively high water flow. Removing wetland vegetation the lines a stream or river leads to poorly anchored soil and an increased water flow, which carries away the soil. The result can be severe erosion and changes to the contours of channels, making them deeper and flatter. As a result, aquatic communities at the erosion location are disrupted or eliminated, and downstream aquatic systems are damaged by silt.

Marsh plant fringes in lakes, estuaries, and oceans protect shorelines from erosion in a similar fashion. The plants reduce soil erosion by binding the soil in their root masses. At the same time, the plants and root masses cushion the force of wave action, retarding scouring of shorelines.

ECONOMIC BENEFITS OF WETLANDS

Appreciation of the economic value of wetlands has undergone a dramatic change since the 1970s. Before that time, wetlands were considered useless, good only for taking up space and breeding mosquitoes. The emphasis was on filling and draining wetlands to turn them into productive land for development and agriculture. In the mid-1970s the growing environmental movement with its emphasis on clean water led to a closer examination of wetlands and their role in watersheds and the global ecosystem. Wetlands are now valued not only for their ecological role but also for their contribution to the economy.

Recreation

Some of the most popular recreational activities, including fishing, hunting, and canoeing, occur in and are dependent on healthy wetlands. The EPA notes in the fact sheet "Economic Benefits of Wetlands" (February 22, 2006, http://www.epa.gov/owow/wetlands/facts/fact4.html) that "more than half of all U.S. adults (98 million people) hunt, fish, birdwatch, or photograph wildlife." In 1991 spending on these activities amounted to $59.5 billion. Figure 7.7 shows that 42% of Americans who observed, fed, or photographed wildlife on trips away from home in 2001 visited wetlands for these activities.

An example of the value of these wetland-related recreational activities can be found in the USFWS's *2001 National Survey of Fishing, Hunting and Wildlife-*

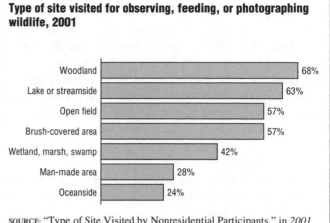

FIGURE 7.7

Type of site visited for observing, feeding, or photographing wildlife, 2001

Site	Percentage
Woodland	68%
Lake or streamside	63%
Open field	57%
Brush-covered area	57%
Wetland, marsh, swamp	42%
Man-made area	28%
Oceanside	24%

SOURCE: "Type of Site Visited by Nonresidential Participants," in *2001 National Survey of Fishing, Hunting, and Wildlife-Associated Recreation*, U.S. Department of the Interior, Fish and Wildlife Service and U.S. Department of Commerce, U.S. Census Bureau, October 2002, http://www.census.gov/prod/2002pubs/FHW01.pdf (accessed February 9, 2007)

Associated Recreation (October 2002, http://www.census .gov/prod/2002pubs/FHW01.pdf). The USFWS states that in 2001, 34.1 million people aged sixteen years and older went fishing and spent an average of $1,046 each; 28.4 million anglers went freshwater fishing and 9.1 million went saltwater fishing. Overall, anglers spent $35.6 billion in 2001 on fishing trips, $4.6 billion on equipment, $6 billion on food and lodging, and $3.5 billion on transportation. They spent nearly $5.3 billion on land-use fees, guide fees, equipment rental, boating expenses, and bait. Camping equipment, binoculars, and special fishing clothing accounted for $721 million in expenditures. Equipment such as boats, vans, and cabins cost $11.6 billion. Anglers spent $3.2 billion on land leasing and ownership and $860 million on magazines, books, membership dues and contributions, licenses, stamps, tags, and permits.

Jerry Leonard reports in *Fishing and Hunting Recruitment and Retention in the U.S. from 1990 to 2005: Addendum to the 2001 National Survey of Fishing, Hunting, and Wildlife-Associated Recreation* (February 2007, http:// library.fws.gov/nat_survey2001_recruitment.pdf) that wetland-related activities are also important to children of all ages. In 2005, 64% of all first-time anglers were aged twenty or younger. In addition, 42% of children of any age living at home were exposed to fishing for the first time in 2005—49% of males and 35% of females.

Commercial Fisheries

The National Marine Fisheries Service notes in *Fisheries of the United States 2005* (February 2007, http:// www.st.nmfs.gov/st1/fus/fus05/fus_2005.pdf) that the value of the U.S. commercial fish landings (the part of the fish catch that is put ashore) in 2005 was $3.9 billion. Nearly 30% of the value of U.S. finfish landings was from species that are dependent on near-coastal waters and their wetlands for breeding and spawning. The EPA estimates in the *National Coastal Condition Report II (2005)* (December 2004, http://www.epa.gov/owow/oceans/nccr/2005/Chapt1_Intro.pdf) that "95% of commercial fish and 85% of sport fish spend a portion of their life cycles in coastal wetland and estuarine habitats. Adult stocks of commercially harvested shrimp, blue crabs, oysters, and other species throughout the United States are directly related to wetland quality and quantity."

Flood Control

Because wetlands function like sponges by absorbing and storing water, they help control flood waters and the resultant loss of life and property. In *Economic Benefits of Wetlands* (May 2006, http://www.epa.gov/OWOW/wetlands/pdf/EconomicBenefits.pdf), the EPA notes that floods in the United States cost about $2 billion annually. However, depending on their size, wetlands can store millions of gallons of water and then release the water slowly after the flood surge has passed, reducing flood damage. Wetlands can also buffer the effects of coastal tropical storms and hurricanes. The EPA suggests that if the Mississippi-Louisiana coastline had more wetland areas, then the effects of Hurricane Katrina would have been lessened.

HISTORY OF WETLANDS USE

Until well into the twentieth century wetlands were considered nature's failure, a waste in nature's economy. For this reason, people sought to increase the usefulness of wetlands. In the agricultural economy of that time, land unable to produce crops or timber was considered worthless. Many Americans began to think of draining these lands, an undertaking that required government funds and resources.

In the nineteenth century state after state passed laws to facilitate drainage of wetlands by the formation of drainage districts and statutes. When a number of landowners in an area petitioned for a drainage project, a hearing was held. A district encompassing the area affected could be created with the power to issue bonds, drain the area, and bill the landholders—petitioners and opponents alike. Coupled with an agricultural boom and technological improvements, reclamation projects multiplied in the late nineteenth and early twentieth centuries. The farmland under drainage doubled between 1905 and 1910 and again between 1910 and 1920. By 1920 state drainage districts in the United States encompassed an area larger than Missouri.

Early Conservationists

The earliest effective resistance came from hunters, sportsmen, and naturalist lobbies. Organizations such as the Izaak Walton League, the Audubon Society, and the American Game Protective Association deplored the destruction by drainage of wildlife habitats and began to press for protection of wetlands. These early conservation efforts met chilly receptions both from the public and the courts. A growing number of Americans, however, were beginning to sympathize with conservationists. Drainage projects were often disappointing—soils had proven to be poorer than expected, and the costs were generally greater than expected.

Reclamation's Failures

Lower Klamath Lake in Northern California became a striking example of reclamation's potential for creating wastelands far more desolate than those they replaced. The lake, a shallow sheet of water fringed by marshes, had been set aside by Theodore Roosevelt in 1908 as a waterfowl sanctuary. Nonetheless, in 1917 the water inflow was cut off to reclaim the land. The lakebed dried up and became prey to dust storms. The peat in the marsh bottom caught fire. Rather than being a reclaimed area of extraordinary fertility, the former wetlands became an ecological travesty. According to the USFWS's Klamath Basin National Wildlife Refuges (April 7, 2007, http://www.fws.gov/klamathbasinrefuges/history.html), even though time has helped reverse the damage, less than 25% of the historic wetland basin remains. In spite of this, the basin continues to support tremendous bird life on a smaller scale.

Efforts were made to help the Klamath Basin recover. According to the news release "President Bush to Propose Record-Level $3.9 Billion for Conservation Programs" (January 30, 2003, http://www.usda.gov/wps/portal/), the U.S. Department of Agriculture (USDA) reports that in the budget for fiscal year 2004, President George W. Bush proposed setting aside $8 million for water conservation and water quality enhancements in the Klamath Basin. The article "Federal Agencies Issue Final Mandates for Klamath Dams" (*California Chronicle*, January 30, 2007) notes that the Departments of Interior and Commerce mandated that fishways and fish ladders be operational in the area, making it economically favorable to remove dams that blocked water to the area, rather than relicensing the dams and providing fishways and ladders using alternative methods. This action led the way toward returning the Klamath River to being a productive salmon river.

Similarly, for many years Florida sought to drain the Everglades, a vast wetland region covering much of the southern part of the state. Efforts there resulted in lands prone to flooding and peat fires. Peat fires are particularly dangerous because they burn underground and can flare up without warning long distances from where they were originally ignited. Costs escalated, and the drainage district went broke. Across the nation the gap between the cost and the value of reclaimed land widened even more. The agricultural depression beginning in the 1920s increased the growing skepticism as to the value of reclamation. Nonetheless, during the Great Depression (1929–39), programs such as the Works Progress Administration and the Reconstruction Finance Corporation encouraged wetland conversion as a way to provide work for many unemployed people. By the end of World War II (1939–45) the total area of drained farmland had increased sharply.

Concern over Property Rights

Dispute over wetlands regulation reflects the nation's ambivalence when private property and public rights intersect, especially because three-fourths of the nation's wetlands are owned by private citizens. In recent years many landowners have complained that wetlands regulation devalued their property by blocking its development. They argued that efforts to preserve the wetlands have gone too far, citing instances where a small wetland precludes the use of large tracts of land. Many people believe this constitutes taking without just compensation.

The "takings" clause of the U.S. Constitution provides that when private property is taken for public use, just compensation must be paid to the owner. Wetland owners claim that when the government, through its laws, eliminates some uses for their land, the value is decreased, and they believe they should be paid for the loss.

In the 1970s and 1980s state courts and the lower federal courts frequently handed down contradictory rulings on the issue of compensation for wetland-related takings. In 1992 the U.S. Supreme Court, in *Lucas v. South Carolina Coastal Council* (505 U.S. 1003), resolved the issue of compensation when land taken for an accepted public good loses significant value.

David Lucas, a homebuilder, bought two residential lots on a South Carolina barrier island in 1986. He planned to build and sell two single-family houses similar to those on nearby lots. At the time he purchased the land, state law allowed house construction on the lots. In 1988 South Carolina passed the Beachfront Management Act to protect the state's beaches from erosion. Lucas's land fell within the area considered in danger of erosion; as a result, Lucas could no longer build the houses.

Lucas went to court, claiming that the Beachfront Management Act had taken his property without just compensation because it no longer had any value if he could not build there. Lucas did not question the right of the state of South Carolina to take his property for the common good. Rather, he claimed the state had to compensate him for the financial loss that resulted from the devaluing of the property.

The Supreme Court said that a state could stop a landowner from building on his or her property only if he or she was using it for a "harmful or noxious" purpose—for example, building a brickyard or a brewery in a residential area. This was not the case. Lucas had planned to build

homes, a legitimate purpose that was neither harmful nor noxious. Although it was possible to define the planned buildings as harmful to South Carolina's ecological resources, this would not be consistent with earlier Court interpretations of "harmful." Only by showing that Lucas had intended to do something "harmful or noxious" with the land could the state take his land without compensation. This the state did not do, and, therefore, it owed him the money.

Invasive Species

People are not the only ones who have dramatically altered wetlands. Nonnative—also called exotic—species have been as devastating to wetlands as humans by changing the nature of the ecosystem, thereby interfering with its function and the survival of native plants and animals. Plants and animals introduced either accidentally or deliberately can cause unexpected harm by displacing native species from their habitat or by placing stress, such as disease or predation, on a native species.

In 1899 the coypu (*Myocastor coypus*) was introduced into California for the fur-farming trade. This animal is a beaverlike aquatic South American rodent that is bred for its fur. This introduction was originally viewed as a way to provide economic benefit. Subsequently, state and federal agencies as well as private interests were responsible for introducing the coypu into the wild in fifteen states to provide a new fur resource. In coastal states such as Maryland and Louisiana, the results have been disastrous.

Coypu live in fresh, intermediate, and brackish marshes and wetlands and feed on the vegetation. They eat all the vegetation in an area, changing a marsh to a barren mudflat. Coypu feed on the base of plant stems and dig for roots and rhizomes in the winter. Their grazing strips large patches of marsh, and their digging turns over the upper peat layer. This conversion of marsh to open water destroys valuable habitat for muskrat, wading birds, amphibians, reptiles, ducks, fish, crabs, and a host of other species, as well as causing erosion and siltation.

Invasive plant species can be as harmful as invasive animal species. Eurasian watermilfoil, phragmites (common reed grass), hydrilla, and purple loosestrife are introduced species that have disrupted wetland systems. Purple loosestrife (*Lythrum salicaria*) is a good example. It is a perennial herb with reddish-purple flowers that may reach six feet in height under the right conditions. It was an important medicinal herb and ornamental as early as two hundred years ago on the East Coast and was probably introduced for this reason. It has no known North American predators and has a high reproductive capacity— up to three hundred thousand seeds per stalk. Because it can outcompete most native wetland plants, it can change the character and ecological function of a marsh. This is a

FIGURE 7.8

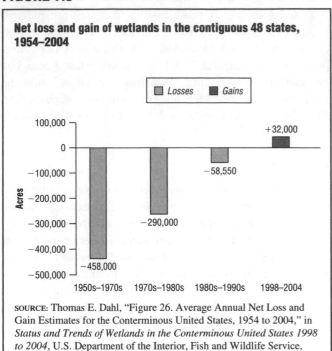

Net loss and gain of wetlands in the contiguous 48 states, 1954–2004

SOURCE: Thomas E. Dahl, "Figure 26. Average Annual Net Loss and Gain Estimates for the Conterminous United States, 1954 to 2004," in *Status and Trends of Wetlands in the Conterminous United States 1998 to 2004*, U.S. Department of the Interior, Fish and Wildlife Service, 2006, http://wetlandsfws.er.usgs.gov/status_trends/national_reports/trends_2005_report.pdf (accessed January 11, 2007)

FIGURE 7.9

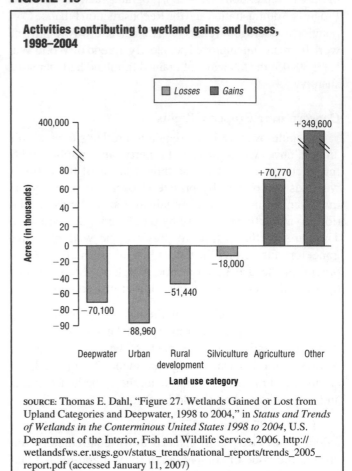

Activities contributing to wetland gains and losses, 1998–2004

SOURCE: Thomas E. Dahl, "Figure 27. Wetlands Gained or Lost from Upland Categories and Deepwater, 1998 to 2004," in *Status and Trends of Wetlands in the Conterminous United States 1998 to 2004*, U.S. Department of the Interior, Fish and Wildlife Service, 2006, http://wetlandsfws.er.usgs.gov/status_trends/national_reports/trends_2005_report.pdf (accessed January 11, 2007)

serious threat because many wetland and other wildlife species are adapted to and depend on specific plants.

LOSS AND GAIN IN WETLAND ACREAGE

When the first Europeans arrived in North America, Thomas E. Dahl and Gregory J. Allord indicate in *History of Wetlands in the Coterminous United States* (March 7, 1997, http://water.usgs.gov/nwsum/WSP2425/history.html) that there were an estimated 221 million acres of wetlands in the lower forty-eight states. In his 2006 report, Dahl notes that in 2004 there were an estimated 107.7 million acres. In the intervening years, more than 50% of the wetlands in the lower forty-eight states had been lost. Wetlands had been drained, dredged, filled, leveled, and flooded to meet human needs. Although natural forces such as erosion, sedimentation, and a rise or drop in sea level may erase wetlands over time, most wetland losses have been caused by humans. Many of the nation's older cities, such as New York City, Baltimore, Philadelphia, New Orleans, and Charleston, are built on filled wetlands.

However, Dahl mentions that after decades of wetland losses there was an annual gain of thirty-two thousand acres of wetlands from 1998 to 2004. (See Figure 7.8.) Additionally, the loss of wetland shrank from 458,000 acres in the 1950s and 1970s to 290,000 acres in the 1970s and 1980s, which is a decrease of 37%. By the 1980s and 1990s the annual wetland loss had declined to 58,500 acres, which is a 79% decrease from the 1970s and 1980s.

Dahl details the reasons for wetlands losses during the 1998–2004 period. Figure 7.9 shows that the highest acreage lost annually was because of urban development. Dahl determines that urban and rural development accounted for an estimated 61% of wetland losses. In addition, 70,100 acres of wetlands were estimated to have been lost to deepwater habitats. Deepwater habitats are "environments where surface water is permanent and often deep, so that water, rather than air, is the principal medium in which the dominant organisms live." Some wetlands (18,000 acres) were lost to silviculture, the planting of trees.

Dahl also discusses the reasons for wetland gains. (See Figure 7.9.) More than 70,000 acres of wetlands were gained from wetland restoration and conservation programs converting agricultural land to wetlands, and nearly 350,000 acres were gained from these programs converting other types of land, such as prairie, forest, or scrub land, to wetlands.

WETLAND PROTECTIVE LEGISLATION AND PROGRAMS

Since the early 1970s conservationists have turned to the courts to challenge reclamation projects and protect wetlands. If drainage once seemed to improve the look of the land, beginning in the 1970s it was more likely to be seen as degrading it. Wetlands turned out to be not wastelands, but systems efficient in harnessing the sun's rays to feed the food chain and play an important role in the global cycle of water, nitrogen, carbon, and sulfur.

As the drainage movement once found support in state laws and federal policies, so did the preservation movement. In 1977 President Jimmy Carter issued an executive order instructing federal agencies to minimize damage to wetlands. In 1989 the EPA adopted a goal of "no net loss" of wetlands, meaning that where a wetland is developed for other uses, the developer must create a wetland elsewhere to maintain an overall constant amount of wetland acreage.

Clean Water Act

Section 404 of the Federal Water Pollution Control Act of 1972 is commonly called the Clean Water Act (CWA). The goal of the CWA is to "restore and maintain the chemical, physical, and biological integrity of the nation's water." Wetlands are considered part of the nation's water and are covered by the CWA.

The CWA authorizes the Army Corps of Engineers to be the primary federal authority for the protection of wetlands. The Corps' jurisdiction encompasses all navigable waters of the United States, plus their tributaries and adjacent wetlands, and includes ocean waters within three nautical miles of the coastline and isolated waters where the use, degradation, or destruction of these waters could affect interstate commerce or foreign commerce. The Corps evaluates the impact of proposed projects that involve wetlands by considering comments from the EPA, the USFWS, the National Marine Fisheries Service, and the affected states. Regulations established under the CWA require that any project affecting more than one-third of an acre of wetlands or five hundred linear feet of streams must be approved by the Corps.

CWA JURISDICTION QUESTIONED. Between 2000 and 2002 legal challenges arose over the extent of the Corps' authority under Section 404 of the CWA and the meaning of certain terms used in the act (such as "waters of the United States" and "navigable waters"). In the CWA information brief *The Supreme Court's SWANCC Decision* (August 2003, http://homer.ornl.gov/nuclearsafety/ nsea/oepa/guidance/cwa/swancc_info_brf.pdf), the U.S. Department of Energy details one such challenge.

According to the brief, the Solid Waste Agency of Northern Cook County (SWANCC) wanted to develop a nonhazardous solid waste disposal facility on a site that contained isolated ponds and wetlands. The Corps denied SWANCC a Section 404 permit to fill those wetlands because they were used by migratory birds. Lower courts found in favor of the Corps, and SWANCC appealed the finding to the U.S. Supreme Court.

On January 9, 2001, the Supreme Court issued the decision *Solid Waste Agency of Northern Cook County v. United States Army Corps of Engineers* (531 U.S. 159). The Court determined that the Corps' authority under the CWA did not extend to isolated wetlands if they were not "adjacent" to navigable waters. It held that the Corps exceeded its statutory authority by asserting CWA jurisdiction over the ponds that SWANCC wanted to fill based solely on the use of those "non-navigable, isolated, intrastate" waters by migratory birds.

PROPOSED CLEAN WATER AUTHORITY RESTORATION ACT. The 2001 SWANCC decision narrowed the scope of wetlands, streams, lakes, and other waters protected under the CWA, which has prompted a move to restore protection. The proposed Clean Water Authority Restoration Act (CWARA) would restore the broad scope of protection to these water bodies and reestablish protections for "isolated" wetlands throughout the United States. The most recent version of the bill was introduced in both the U.S. House of Representatives and the U.S. Senate in May 2005. In early 2007 the bill was in the first stages of the legislative process in both chambers of Congress, having been referred to subcommittees for consideration.

Farm Bill of 1996

The 1996 Farm Bill reauthorized the Conservation Reserve Program and created the Wetlands Reserve Program. The two programs are designed to protect and restore wetlands.

WETLANDS RESERVE PROGRAM. The Wetlands Reserve Program (WRP) is a voluntary USDA program and has been implemented in forty-nine states. The program provides farmers with financial incentives, such as a fair market price for land, to retire marginal farmland and, in many cases, to restore and protect wetlands. In the key points sheet "Farm Bill 2002" (September 2004, http://www.nrcs.usda.gov/ programs/farmbill/2002/pdf/WRPKyPts.pdf), the USDA notes that the program had enrolled nearly 1.5 million acres in 2004. Retiring cropland through the WRP has benefited the recovery of threatened or endangered species and has protected wetlands. The WRP was reauthorized under the Farm Bill of 2002 and is to extend through December 31, 2007.

CONSERVATION RESERVE PROGRAM. The Conservation Reserve Program (CRP) was originally authorized in the Farm Bill of 1985 as a soil conservation strategy that included paying farmers to retire marginal cropland from production for ten years. Its political support came from its potential to reduce expensive crop surpluses. The Natural Resources Conservation Service notes in "Conservation Reserve Program" (March 26, 2007, http://www.nrcs.usda.

gov/Programs/crp/) that under the CRP the Farm Service Agency pays farmers to plant natural vegetation in their "highly erodible cropland or other environmentally sensitive acreage."

Unlike the WRP, farmers do not have to permanently retire their land under this program, but instead can do so for ten-year intervals. As a wetland protection and restoration strategy, the program has been successful in terms of the thousands of acres of cropland that have been restored to a natural state, which in many cases includes wetlands. The CRP was reauthorized in the Farm Bill of 2002, extending the program through December 31, 2007.

State Wetland Protection Programs

Many states have enacted their own state laws to protect wetlands. These laws may complement or be more stringent than federal regulations. For example, Maryland has had state laws to protect tidal wetlands since the early 1970s. In 1989 Maryland adopted its Nontidal Wetlands Act to provide the same protections to freshwater wetlands.

Besides using their CWA authority, states have included wetland protection in their water quality standards, passed laws protecting ecologically important wetlands such as the Dismal Swamp in Virginia and North Carolina, established mitigation banking, and created public education programs to increase public awareness of the value of wetlands. Several states have set up special funds to buy important wetlands.

WETLAND GAINS

As shown in Figure 7.8, wetland gains have been made in recent years. Figure 7.9 shows the reasons for these gains. Many efforts are ongoing at the private, local, state, and federal levels to protect existing wetlands and to create new ones. Wetland losses can be offset by restoring, creating, enhancing, replacing, or reallocating wetlands:

- Wetland restoration—the return of a wetland to a close approximation of its condition before disturbance, including reestablishment of its predisturbance aquatic functions and related physical, chemical, and biological characteristics.

- Creation—the construction of a wetland in an area that was not a wetland within the past one hundred to two hundred years and is isolated from other wetlands.

- Enhancement—the modification of one or more structural features of an existing wetland to increase one or more functions based on management objectives. Enhancement, while causing a positive gain in one function, frequently results in a reduction in another function.

- Replacement or reallocation—activities in which most or all of an existing wetland is converted to a

different type of wetland and has the same drawback as enhancement.

Each of these approaches has benefits and drawbacks.

Private Initiatives

Many of the wetland areas in the United States are privately owned. A number of government programs, both regulatory and voluntary, exist to foster wetland protection, and some foster both restoration and enhancement. Some of the most successful wetland programs and projects are the result of private initiatives. Frequently, private organizations form partnerships with landowners to buy, lease, or create easements paid for with private, or a mix of private and public, funds.

Organizations such as the Nature Conservancy, Ducks Unlimited (DU), the Audubon Society, the Chesapeake Bay Foundation, and hundreds of others are working with private landowners, corporations, local communities, volunteers, and federal and state agencies in innovative projects to protect and restore wetlands. For example, the Nature Conservancy oversees many wetland restoration projects, including two on the Illinois River—Spunky Bottoms (2007, http://www.nature.org/initiatives/freshwater/work/illinoisriver.html) and Emiquon (2007, http://www.nature.org/wherewework/northamerica/states/illinois/preserves/art1112.html)—that aim to return more than eighty-five hundred acres of farmed land to their original wetland state.

In another example, the DU is working with the National Resources Conservation Service to implement the WRP in the Mississippi Alluvial Valley, which according to the DU, in "Mississippi Alluvial Valley" (2007, http://www.ducks.org/conservation/initiative21.aspx), historically comprised 24.7 million acres of hardwood bottom stretching from southern Illinois to Louisiana. In "Conservation in Mississippi" (2007, http://www.ducks.org/Page1666.aspx), the DU reports that it is working to conserve over 250,000 acres of waterfowl habitat throughout Mississippi by restoring hydrology and planting bottomland hardwood seedlings.

Constructed Wetlands

Constructed wetlands are marshes that are built to filter contaminated water. They consist of soil and drainage materials (such as gravel), water, plants, and microorganisms. Using constructed wetlands for wastewater treatment is a simple, economical, and environmentally friendly method that is being used more frequently than in the past.

Constructed wetland treatment systems are designed and built to use the natural processes involving wetland soils, vegetation, and their associated microbes to help treat wastewater. They are designed to take advantage of many of the same processes that occur in wetlands but in a more controlled manner. Even though some of these systems are operated solely to treat wastewater, others are designed with the multiple objectives of using treated wastewater as a source of water for the creation or restoration of wetland habitat for wildlife and environmental enhancement. The primary drawback to constructed wetlands for wastewater treatment is that they are land intensive; large land tracts are not always available at affordable prices.

There are two general types of constructed wetland treatments: subsurface flow systems and free water surface systems. Both types are usually built in basins or channels with a natural or human-made subsurface barrier to limit seepage. The subsurface flow systems keep water flowing through soil, sand, gravel, or crushed rock underground to minimize odors and other related problems. (See Figure 7.10.) Subsurface flow systems are also known as rock-reed filters, vegetated submerged bed systems, and root-zone systems. Free water surface systems are designed to simulate natural wetlands, with the water flowing over the soil surface at shallow depths.

The EPA's Office of Water reports in *Constructed Treatment Wetlands* (August 2004, http://www.epa.gov/owow/wetlands/pdf/ConstructedW.pdf) that approximately five thousand constructed wetland treatment systems have been built in Europe and about one thousand are operating in the United States.

Marsh construction and wetland rehabilitation as a method of disposing of dredged materials are another growing source of wetland construction. The Army Corps of Engineers has been using dredged material to restore or construct marshes since 1969. Dredged material is placed on shallow bay bottoms to build up elevations to an intertidal level, usually by pumping dredged material to the marsh construction site. If the site is exposed to high wind or wave action, protective structures such as rock or concrete breakwaters are built. Vegetation can be planted or the site may be left to develop naturally. Generally, within two to three years, these sites are indistinguishable from natural wetlands in appearance.

Restoration of the Florida Everglades

The Everglades is a premier wetland in the United States. It is designated as an International Biosphere Reserve, a World Heritage Site, and a Wetland of International Importance. According to the World Heritage Committee, the Everglades is the only U.S. site on the List of World Heritage in Danger (April 10, 2007, http://whc.unesco.org/en/danger/). Figure 7.11 shows the location of the Everglades and how it has been reduced to about half its former size.

According to the South Florida Ecosystem Restoration Task Force (April 6, 2007, http://www.sfrestore.org/), the Everglades is part of the South Florida Ecosystem, an eighteen-thousand-square-mile region extending from the

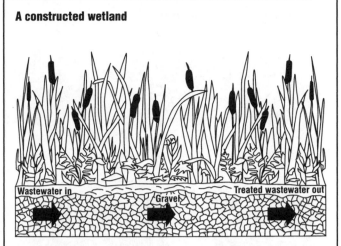

FIGURE 7.10

A constructed wetland

Notes: Marsh plants (cattails, reeds, etc.) are grown in beds of soil or gravel through which wastewater flows. Wetlands are useful to further treat wastewater from a lagoon. This is a low-cost system that needs minimal attention from an operator. Periodically, plants need to be checked and sometimes harvested at the end of the growing season. The system requires relatively less land than many land treatment systems. The system may be operated year-round in most climates.

SOURCE: "Figure 2.3. Constructed Wetland," in *Water Pollution: Information on the Use of Alternative Wastewater Treatment Systems*, U.S. General Accounting Office, September 1994, http://archive.gao .gov/t2pbat2/152794.pdf (accessed January 11, 2007)

Kissimmee River near Orlando to the Florida Keys. Originally a wide expanse of wetland, pine forests, mangroves, coastal islands, and coral reefs, in the twenty-first century it is one of the nation's most highly populated and manipulated regions. Its freshwater supply comes from rainfall in the Kissimmee River Basin and southward, mostly in May through October.

Slow and rain driven, the natural cycle of freshwater circulation feeding the Everglades historically built up in shallow Lake Okeechobee, which averages twelve feet deep and covers about 730 square miles. Thus began the flow of the wide, shallow "river of grass," as it was called by Native Americans. Fifty miles wide in places, one to three feet deep in the slough's center, and only six inches deep elsewhere, it flowed south at a rate of about one hundred feet per day across the saw grass of the Everglades to the mangrove estuaries on the Gulf of Mexico. A six-month dry season followed this flow. During the dry season water levels gradually drop. The plants and animals of the Everglades are adapted to the alternating wet and dry seasons.

During the past one hundred years an elaborate system of dikes, canals, levees, floodgates, and pumps was built to move water to agricultural fields, urban areas, and the Everglades National Park. Water runoff from agriculture and urban development brought excess nutrients into the Everglades, reducing production of beneficial algae and promoting unnatural growth of other vegetation. Ill-timed human manipulation of the water supply interfered

FIGURE 7.11

The Everglades, past and present

Location of the Everglades.

☐ *Natural areas of the Everglades*

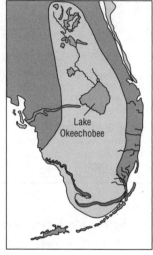 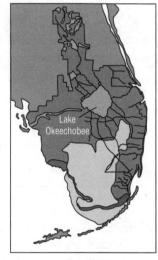

| Past | Present |

Historically, the natural areas of the Everglades extended well north of Lake Okeechobee and south to Florida Bay and the Gulf of Mexico.

Today, the bulk of the natural areas remaining in the ecosystem primarily include the Everglades National Park and Big Cypress National Preserve, as well as state water conservation areas.*

*Other smaller natural areas are dispersed throughout southern Florida, such as national wildlife refuges and state, local, or privately owned lands, but are not shown in the figure.

SOURCE: South Florida Water Management District, "Figure 23. The Everglades—Past and Present," in *Freshwater Supply: States' Views of How Federal Agencies Could Help Them Meet the Challenges of Expected Shortages*, U.S. General Accounting Office, July 2003, http://www.gao.gov/new.items/d03514.pdf (accessed February 10, 2007)

with the natural water cycle, ruining critical spawning, feeding, and nesting conditions for many species.

The Florida legislature has enacted a number of laws to combat the growing water shortage in Florida, includ-ing the Everglades. The 1981 Save Our Rivers Act and the 1990 Preservation 2000 Fund authorized the water management districts to buy property to protect water sources, groundwater recharge, and other natural resources. The South Florida Water Management District (SFWMD; 2007, http://www.evergladesplan.org/pm/progr_land_aqui sition.cfm), an agency that oversees flood protection and water supply, began buying out landowners in the eastern Everglades area in hopes of retaking thousands of acres of agricultural and residential property at an estimated cost of $2.2 billion. The action is aimed at restoring water flow to Everglades National Park.

In 1998 the Army Corps of Engineers and the SFWMD released their plan for improving Florida's ecological and economic health: the Comprehensive Everglades Restora-tion Plan (CERP). This plan covers the entire region and its water problems and focuses on recovering the major char-acteristics that defined the "river of grass." Specifically, the plan calls for:

• Reducing the freshwater flows into the Caloosa-hatchee River and the St. Lucie Canal, thereby restor-ing to the Everglades water now lost to the tide.

• Returning the water flow in the Kissimmee River to its former floodplain to achieve a more meandering river system.

• Restoring forty thousand acres of marshes for water storage and filtration to remove nutrients before water entering the Everglades.

• Modifying water deliveries through improved timing and distribution to mimic historic water conditions.

• Reestablishing historic flows and water levels to sloughs feeding into Florida Bay to restore natural estuarine salinity.

The Water Resources Development Act of 2000 approved the CERP. The SFWMD (2007, http://www.ever gladesplan.org/about/about_cerp_brief.aspx) states that this plan will take more than thirty years to carry out and will cost an estimated $7.8 billion. The primary goal of the project is to restore critical water flows to the Everglades and ensure adequate water supplies for cities, communities, and farmers in southern Florida well into the future. The cost of the project will be shared equally between the state of Florida and the federal government.

CHAPTER 8
THE ARID WEST—WHERE WATER IS SCARCE

CLIMATE OF THE AMERICAN WEST

The United States is a nation relatively rich in water resources. According to Susan S. Hutson et al., in *Estimated Use of Water in the United States in 2000* (2004, http://pubs.usgs.gov/circ/2004/circ1268/pdf/circular1268.pdf), in the lower forty-eight states the total supply of water is about 1.4 trillion gallons per day. Nevertheless, while the nation as a whole is water-rich, this abundance is not spread evenly throughout the country. Some areas have more water than others, whereas some have a higher need than others. Those with the greatest need do not always have adequate water resources, a situation that can lead to serious problems and conflicts.

Much of the U.S. West is arid (characterized by desert land) and semiarid (prairie land), with limited and inconsistent supplies of water. The West includes Washington, Oregon, and California (the Pacific states); and Idaho, Montana, Wyoming, Colorado, Utah, Nevada, New Mexico, and Arizona (the Mountain states). From the Rocky Mountains, which form the Continental Divide, to the shores of California, lay the dry basins and deserts of this vast western region of the country.

Alfred de Grazia notes in *The American Way of Government* (1957) that the lower forty-eight states comprise 1.9 billion acres of land. Almost half of this land is semiarid and arid and receives less than twenty inches of precipitation per year. The small amount of rain and snow that falls is unevenly distributed. For example, the Weather Channel (April 2007, http://www.weather.com/) reports that Flagstaff, Arizona, receives more than twenty inches of precipitation a year, but in Phoenix and Tucson, where most of Arizona's people live and most of the agriculture is located, the yearly rainfall averages barely nine inches. The reason for this is the North Pacific high pressure system. This is a huge zone of high atmospheric pressure that is the characteristic weather pattern for the Pacific Ocean off the coast of North America. The North Pacific High pushes most precipitation toward the north.

No resource is as vital to the West's urban centers, agriculture, industry, recreation, scenic beauty, and environmental preservation as water. Throughout the history of the West, especially in California, battles have raged over who gets how much of this precious resource. The fundamental controversy is one of distribution, combined with conflicts between competing interests over the use of available supplies.

POPULATION GROWTH IN THE WEST

Although the climate is arid and semiarid in much of the West, and water is a necessary and less available resource than in other parts of the United States, the population of the West is booming. According to data from the U.S. Census Bureau, population growth exploded in the West between 1990 and 2000. Figure 8.1 shows that from 2000 to 2006, growth in the United States was focused in the West as well as in the South. The maps show that in terms of numbers of people, states in the West and the South had the greatest rise. In terms of percent change, two western states—Nevada and Arizona—had the highest growth rate of all states from 2000 to 2006.

The Census Bureau projects that the two highest percentages of population growth from 2000 to 2030 will be in the West (45.8%) and the South (42.9%). (See Table 8.1.) The top five states with the projected highest percentage gains will be Nevada (114.3%), Arizona (108.8%), Florida (79.5), Texas (59.8%), and Idaho (52.2%).

Table 8.2 shows population growth in metropolitan areas (at least one urbanized area of 50,000 or more people), micropolitan areas (at least one urbanized area of 10,000 to 49,999 people), and areas outside either (often called rural areas). The highest growth rate in metropolitan areas from 1990 to 2000 was in the Mountain states of the

FIGURE 8.1

Numerical and percent change in population for states and Puerto Rico, July 1, 2005–July 1, 2006 and 2000–06

Numerical change

Numerical change in population		
■ 200,000 to 579,275	■ 60,000 to 199,999	■ 10,000 to 59,999
□ 0 to 9,999	□ −9,999 to −1	□ −219,563 (LA)

July 2005 to July 2006

Numerical change in population		
■ 500,000 to 2,655,993	■ 200,000 to 499,999	■ 60,000 to 199,999
□ 0 to 59,999	□ −6,333 (ND)	□ −181,190 (LA)

2000 to 2006

West, at 36.7%. From 2000 to 2003 the Mountain states of the West had the highest growth rate again: 7.8%.

SOURCES OF WESTERN WATER SUPPLIES

Precipitation (rain, snow, and sleet) is the main source of essentially all freshwater supplies, largely con-

FIGURE 8.1

Numerical and percent change in population for states and Puerto Rico, July 1, 2005–July 1, 2006 and 2000–06 [CONTINUED]

Percent change

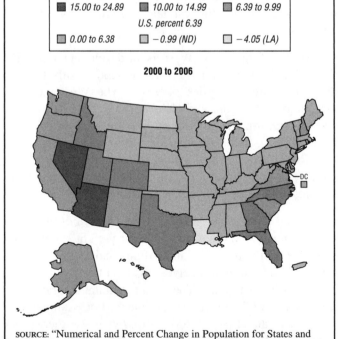

Percentage change in population		
■ 2.00 to 3.58	■ 1.50 to 1.99	■ 0.98 to 1.49
	U.S. percent 0.98	
□ 0.00 to 0.97	□ −0.56 to-0.01	□ −4.87 (LA)

July 2005 to July 2006

Percentage change in population		
■ 15.00 to 24.89	■ 10.00 to 14.99	■ 6.39 to 9.99
	U.S. percent 6.39	
□ 0.00 to 6.38	□ −0.99 (ND)	□ −4.05 (LA)

2000 to 2006

SOURCE: "Numerical and Percent Change in Population for States and Puerto Rico: July 1, 2005 to July 1, 2006 and 2000 to 2006," in *Population Estimates: Maps, States"* U.S. Census Bureau, Population Division, 2006, http://www.census.gov/popest/gallery/maps/StEst_2006_4up_2.pdf (accessed January 11, 2007)

trolling the availability of surface water and groundwater. In the arid regions of the West, much of the available

TABLE 8.1

Change in total population, numerical and percent, for regions, divisions, and states, 2000–30

Region, division, and state	Numerical change 2000 to 2010	Numerical change 2010 to 2020	Numerical change 2020 to 2030	Numerical change 2000 to 2030	Percent change 2000 to 2010	Percent change 2010 to 2020	Percent change 2020 to 2030	Percent change 2000 to 2030
United States	27,513,675	26,868,965	27,779,889	82,162,529	9.8	8.7	8.3	29.2
Northeast	2,190,801	1,350,258	535,631	4,076,690	4.1	2.4	0.9	7.6
New England	816,272	570,739	313,487	1,700,498	5.9	3.9	2.0	12.2
Maine	82,211	51,531	2,432	136,174	6.4	3.8	0.2	10.7
New Hampshire	149,774	139,191	121,720	410,685	12.1	10.0	8.0	33.2
Vermont	43,685	38,174	21,181	103,040	7.2	5.9	3.1	16.9
Massachusetts	300,344	206,105	156,463	662,912	4.7	3.1	2.3	10.4
Rhode Island	68,333	37,578	−1,289	104,622	6.5	3.4	−0.1	10.0
Connecticut	171,925	98,160	12,980	283,065	5.0	2.7	0.4	8.3
Middle Atlantic	1,374,529	779,519	222,144	2,376,192	3.5	1.9	0.5	6.0
New York	467,215	133,248	−99,491	500,972	2.5	0.7	−0.5	2.6
New Jersey	603,881	443,404	340,805	1,388,090	7.2	4.9	3.6	16.5
Pennsylvania	303,433	202,867	−19,170	487,130	2.5	1.6	−0.1	4.0
Midwest	2,998,657	2,063,742	1,042,123	6,104,522	4.7	3.1	1.5	9.5
East North Central	1,886,286	1,167,410	429,731	3,483,427	4.2	2.5	0.9	7.7
Ohio	223,041	67,877	−93,530	197,388	2.0	0.6	−0.8	1.7
Indiana	311,654	234,869	183,100	729,623	5.1	3.7	2.8	12.0
Illinois	497,601	319,826	196,172	1,013,599	4.0	2.5	1.5	8.2
Michigan	490,239	267,310	−1,821	755,728	4.9	2.6	0.0	7.6
Wisconsin	363,751	277,528	145,810	787,089	6.8	4.8	2.4	14.7
West North Central	1,112,371	896,332	612,392	2,621,095	5.8	4.4	2.9	13.6
Minnesota	501,157	480,133	405,361	1,386,651	10.2	8.9	6.9	28.2
Iowa	83,583	10,589	−65,324	28,848	2.9	0.4	−2.2	1.0
Missouri	326,867	277,804	230,291	834,962	5.8	4.7	3.7	14.9
North Dakota	−5,577	−6,511	−23,546	−35,634	−0.9	−1.0	−3.7	−5.5
South Dakota	31,555	15,540	−1,477	45,618	4.2	2.0	−0.2	6.0
Nebraska	57,734	33,681	17,569	108,984	3.4	1.9	1.0	6.4
Kansas	117,052	85,096	49,518	251,666	4.4	3.0	1.7	9.4
South	13,346,794	13,987,205	15,698,518	43,032,517	13.3	12.3	12.3	42.9
South Atlantic	8,022,621	8,650,245	9,651,190	26,324,056	15.5	14.5	14.1	50.8
Delaware	100,742	78,867	49,449	229,058	12.9	8.9	5.1	29.2
Maryland	608,484	592,656	524,625	1,725,765	11.5	10.0	8.1	32.6
District of Columbia	−42,274	−49,245	−47,126	−138,645	−7.4	−9.3	−9.8	−24.2
Virginia	931,730	907,150	907,624	2,746,504	13.2	11.3	10.2	38.8
West Virginia	20,797	−28,029	−81,153	−88,385	1.2	−1.5	−4.5	−4.9
North Carolina	1,296,510	1,363,466	1,518,450	4,178,426	16.1	14.6	14.2	51.9
South Carolina	434,692	375,873	325,992	1,136,557	10.8	8.5	6.8	28.3
Georgia	1,402,627	1,254,673	1,174,085	3,831,385	17.1	13.1	10.8	46.8
Florida	3,269,313	4,154,834	5,279,244	12,703,391	20.5	21.6	22.6	79.5
East South Central	1,040,901	915,117	923,457	2,879,475	6.1	5.1	4.9	16.9
Kentucky	223,348	159,314	130,567	513,229	5.5	3.7	3.0	12.7
Tennessee	541,569	549,818	599,964	1,691,351	9.5	8.8	8.8	29.7
Alabama	149,230	132,585	145,328	427,143	3.4	2.9	3.1	9.6
Mississippi	126,754	73,400	47,598	247,752	4.5	2.5	1.6	8.7
West South Central	4,283,272	4,421,843	5,123,871	13,828,986	13.6	12.4	12.8	44.0
Arkansas	201,639	185,180	179,989	566,808	7.5	6.4	5.9	21.2
Louisiana	143,703	106,481	83,473	333,657	3.2	2.3	1.8	7.5
Oklahoma	140,862	144,174	177,561	462,597	4.1	4.0	4.8	13.4
Texas	3,797,068	3,986,008	4,682,848	12,465,924	18.2	16.2	16.4	59.8
West	8,977,423	9,467,760	10,503,617	28,948,800	14.2	13.1	12.9	45.8
Mountain	3,568,184	3,816,570	4,352,383	11,737,137	19.6	17.6	17.0	64.6
Montana	66,403	54,137	22,163	142,703	7.4	5.6	2.2	15.8
Idaho	223,338	224,042	228,291	675,671	17.3	14.8	13.1	52.2
Wyoming	26,104	11,062	−7,969	29,197	5.3	2.1	−1.5	5.9
Colorado	530,293	447,313	513,490	1,491,096	12.3	9.3	9.7	34.7
New Mexico	161,179	104,116	15,367	280,662	8.9	5.3	0.7	15.4
Arizona	1,506,749	1,819,067	2,255,949	5,581,765	29.4	27.4	26.7	108.8
Utah	361,844	395,081	495,273	1,252,198	16.2	15.2	16.6	56.1
Nevada	692,274	761,752	829,819	2,283,845	34.6	28.3	24.0	

precipitation evaporates shortly after the rains stop. Tucson, Arizona, for example, receives most of its annual rainfall from heavy thunderstorms during the hottest months of the year—between July and September—when much of that rainfall is lost through evaporation.

Runoff refers to water that is not immediately absorbed into the ground during a rain and runs off into lower-lying areas or surrounding lakes and streams. Runoff is the primary measure of a region's water supply. Besides rain, a large share of the West's runoff comes

TABLE 8.1

Change in total population, numerical and percent, for regions, divisions, and states, 2000–30 [CONTINUED]

Region, division, and state	Numerical change 2000 to 2010	Numerical change 2010 to 2020	Numerical change 2020 to 2030	Numerical change 2000 to 2030	Percent change 2000 to 2010	Percent change 2010 to 2020	Percent change 2020 to 2030	Percent change 2000 to 2030
Pacific	5,409,239	5,651,190	6,151,234	17,211,663	12.0	11.2	11.0	38.2
Washington	647,842	890,173	1,192,665	2,730,680	11.0	13.6	16.0	46.3
Oregon	369,597	469,397	573,525	1,412,519	10.8	12.4	13.5	41.3
California	4,195,486	4,139,609	4,238,118	12,573,213	12.4	10.9	10.0	37.1
Alaska	67,177	80,312	93,253	240,742	10.7	11.6	12.0	38.4
Hawaii	129,137	71,699	53,673	254,509	10.7	5.3	3.8	21.0

SOURCE: Table 7. Interim Projections: Change in Total Population for Regions, Divisions, and States: 2000 to 2030, in *Interim State Population Projections, 2005*, U.S. Census Bureau, Population Division, April 21, 2005, http://www.census.gov/population/projections/PressTab7.xls (accessed February 10, 2007)

TABLE 8.2

Population change in metropolitan areas in the U.S., 1990–2000 and 2000–03

[Numerical change in thousands. Data are for April 1, 1990; April 1, 2000, estimates base; and July 1, 2003.]

Geographic area	2000–2003 Total	2000–2003 Metropolitan	2000–2003 Micropolitan	2000–2003 Outside CBSAs	1990–2000 Total	1990–2000 Metropolitan	1990–2000 Micropolitan	1990–2000 Outside CBSAs
Numerical change								
United States	9,387	8,815	475	97	32,713	28,641	2,667	1,406
Regions and divisions								
Northeast region	805	739	51	15	2,786	2,590	137	59
New England division	283	238	32	13	716	624	64	28
Middle Atlantic division	522	501	19	2	2,070	1,966	73	31
Midwest region	1,011	980	51	−20	4,726	3,945	510	271
East north central division	682	624	37	20	3,146	2,613	314	219
West north central division	329	355	14	−40	1,580	1,331	197	52
South region	4,304	4,000	233	71	14,789	12,719	1,314	756
South Atlantic Division	2,578	2,359	158	61	8,200	7,165	683	352
East south central division	318	267	46	5	1,847	1,287	361	198
West south central division	1,407	1,375	28	5	4,742	4,268	269	205
West region	3,267	3,096	139	32	10,413	9,387	706	320
Mountain division	1,211	1,125	67	19	4,514	3,861	428	226
Pacific division	2,056	1,971	72	13	5,898	5,526	278	94
Percent change								
United States	3.3	3.8	1.6	0.5	13.2	14.0	10.0	7.8
Regions and divisions								
Northeast region	1.5	1.5	1.3	1.0	5.5	5.7	3.7	4.3
New England division	2.0	2.0	2.9	2.1	5.4	5.4	6.1	4.8
Middle Atlantic division	1.3	1.4	0.7	0.2	5.5	5.8	2.8	4.0
Midwest region	1.6	2.0	0.6	−0.3	7.9	8.8	6.0	4.3
East north central division	1.5	1.7	0.7	0.7	7.5	7.7	5.9	7.8
West north central division	1.7	2.9	0.4	−1.1	8.9	12.1	6.2	1.5
South region	4.3	5.1	1.9	0.8	17.3	19.2	12.0	9.0
South Atlantic division	5.0	5.5	3.1	1.7	18.8	19.9	15.5	11.2
East south central division	1.9	2.5	1.3	0.2	12.2	13.6	11.4	7.7
West south central division	4.5	5.5	0.8	0.2	17.8	20.6	8.1	7.7
West region	5.2	5.5	3.2	1.4	19.7	19.9	19.1	16.2
Mountain division	6.7	7.8	3.0	1.3	33.1	36.7	23.3	17.2
Pacific division	4.6	4.7	3.4	1.6	15.1	15.1	15.0	14.0

Note: Metropolitan areas contain at least one urbanized area of 50,000 or more people. Micropolitan areas contain at least one urbanized area of 49,999 people. Together these areas are called core based statistical areas (CBSAs). Territory not included as either is called "outside CBSAs."

SOURCE: "Table 2. Population Change by Core Based Statistical Area (CBSA) Status for the United States, Regions, and Divisions: 1990–2000 and 2000–2003," in *Population Change in Metropolitan and Micropolitan Statistical Areas: 1990–2003*, U.S. Census Bureau, September 2005, http://www.census.gov/prod/2005pubs/p25-1134.pdf (accessed February 15, 2007)

from the melting of mountain snowpacks, which are essentially huge reservoirs of frozen water that slowly release their supplies during the spring and summer. Much of western agriculture depends on this meltwater becoming available during the growing season.

Changes in climate can adversely affect the volume of snowpacks. The U.S. Global Change Research Program (USGCRP) projects that the snowpack of the West's mountain ranges will likely decrease as the climate continues to warm, despite a projected increase in

FIGURE 8.2

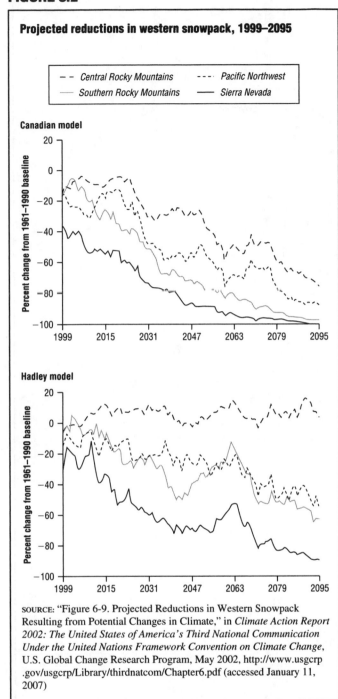

Projected reductions in western snowpack, 1999–2095

Central Rocky Mountains
Pacific Northwest
Southern Rocky Mountains
Sierra Nevada

Canadian model

Hadley model

SOURCE: "Figure 6-9. Projected Reductions in Western Snowpack Resulting from Potential Changes in Climate," in *Climate Action Report 2002: The United States of America's Third National Communication Under the United Nations Framework Convention on Climate Change*, U.S. Global Change Research Program, May 2002, http://www.usgcrp.gov/usgcrp/Library/thirdnatcom/Chapter6.pdf (accessed January 11, 2007)

precipitation. (See Figure 8.2.) The USGCRP suggests that in the coming years more precipitation will fall as rain (rather than as snow) and that snowpack will develop later and melt earlier. As a result, peak stream flows will likely come earlier in the spring, and summer flows will be reduced. The change in the timing of runoff from snowmelt is likely to have implications for water management, flood protection, irrigation, and planning.

Surface Water

The Colorado River is a major source of water for many western states: Colorado, New Mexico, Utah, Wyoming,

Arizona, California, and Nevada. (See Figure 8.3.) The Colorado River Basin is the land area that drains into the Colorado River and its tributaries. In the *2005 Review: Water Quality Standards for Salinity, Colorado River System* (October 2005, http://crb.ca.gov/Salinity/2005/2005%20Triennial%20Review.pdf), the Colorado River Basin Salinity Control Forum reports that this river supplies water to about 7.5 million people within the basin area and another 25.4 million people outside the basin area in the states mentioned previously. The forum states, "About 2.3 million acres are irrigated within the Colorado River Basin and hundreds of thousands of additional acres are irrigated by waters exported from the Colorado River Basin. Hydroelectric power facilities along the Colorado River and its tributaries generate approximately 12 billion kilowatt-hours annually which is used both inside and outside of the Colorado River Basin."

The Colorado River is managed and operated under guidelines called the Law of the River. This law is actually a collection of federal laws, state compacts, court decisions, and regulatory guidelines. Under the Law of the River, the Colorado River is apportioned among the seven basin states and Mexico, into which the Colorado River flows. (See Figure 8.3.) The legal right for a state to use a certain amount of water from the Colorado River over a given period is called a water entitlement. Aqueducts, which are artificial water channels, bring water from the river to the general area where it is needed. It is distributed to users from that point.

Along with the Colorado River, another major source of water for California are the Sacramento and San Joaquin rivers. These rivers come together at the Sacramento–San Joaquin Delta and then flow into the northern arm of San Francisco Bay. The rivers, the delta, and their location within the state are shown in Figure 8.4. Jay Lund et al. note in *Envisioning Futures for the Sacramento–San Joaquin Delta* (2007, http://www.ppic.org/content/pubs/report/R_207JLR.pdf) that "the Delta is considered the hub of the state's water supply because it is used as a transit point for this water."

The term *transit point* suggests that the water is traveling somewhere, and it is. In the southern end of the delta near the city of Tracy, the California Aqueduct begins. (See Figure 8.4.) This 444-mile-long artificial channel carries water from Northern to Southern California. Much of the time the water flows by gravity, but pumps are used in certain places along the way.

Groundwater

The other source of water in the West is groundwater (underground supplies). Most groundwater is found in aquifers, underground saturated zones full of water within the spaces between soil and rocks, and within the rocks themselves. These saturated zones are recharged

FIGURE 8.3

Colorado River and river basin states

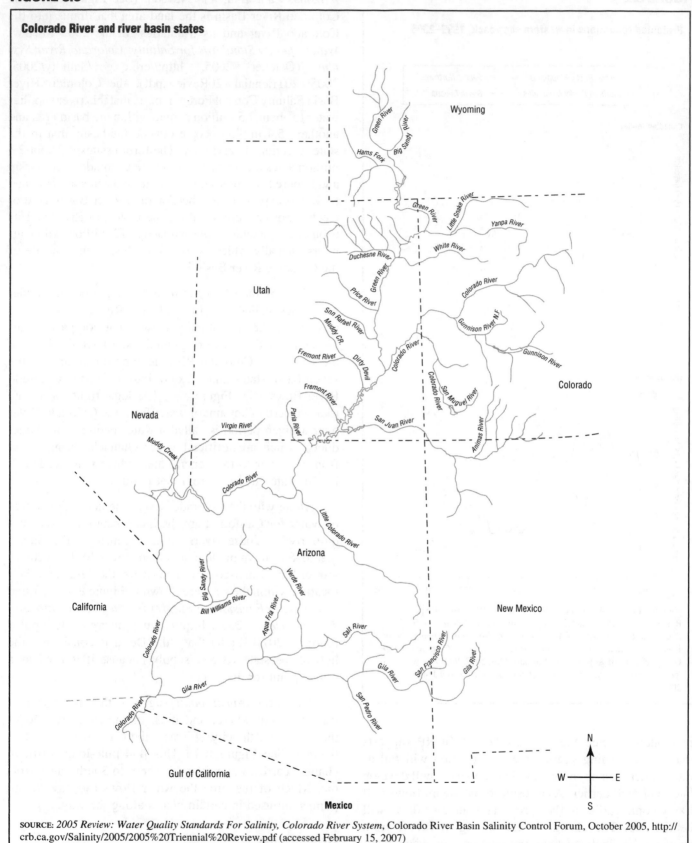

SOURCE: *2005 Review: Water Quality Standards For Salinity, Colorado River System*, Colorado River Basin Salinity Control Forum, October 2005, http://crb.ca.gov/Salinity/2005/2005%20Triennial%20Review.pdf (accessed February 15, 2007)

(replenished) primarily from rainfall percolating through the soils. Water from streams, lakes, wetlands, and other water bodies may also seep into the saturated zones. In the saturated zone, water is under pressure that is higher than atmospheric pressure. When a well is dug into the saturated zone, water flows from the area of higher pressure

FIGURE 8.4

The Sacramento–San Joaquin Delta

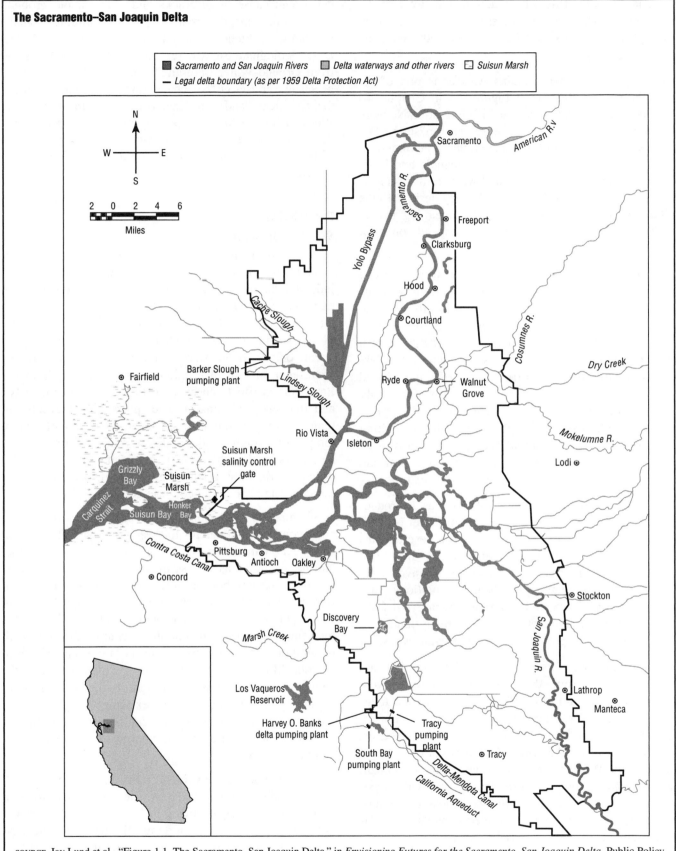

■ Sacramento and San Joaquin Rivers　　■ Delta waterways and other rivers　　□ Suisun Marsh
— Legal delta boundary (as per 1959 Delta Protection Act)

SOURCE: Jay Lund et al., "Figure 1.1. The Sacramento–San Joaquin Delta," in *Envisioning Futures for the Sacramento–San Joaquin Delta*, Public Policy Institute of California, 2007, http://www.ppic.org/content/pubs/report/R_207JLR.pdf (accessed February 16, 2007)

(in the ground) to the area of lower pressure (in the hollow well), and the well fills with water to the level of the existing groundwater. If the pressure is strong enough, the water will flow freely to the surface; otherwise the water must be pumped.

The Ogallala or High Plains Aquifer is one of the world's largest aquifers. (See Figure 4.4 in Chapter 4.) According to the High Plains Aquifer Information Network (2007, http://www.hiplain.org/states/index.cfm?state=9&c=1&c=84), it covers 174,000 square miles stretching from southern South Dakota to the Texas panhandle and is the largest single source of underground water in the country. The U.S. Environmental Protection Agency designates the Ogallala Aquifer a sole-source aquifer, meaning that at least 50% of the population in the area depends on it for its water supply.

The Ogallala Aquifer provides water to portions of eight western and midwestern states: Colorado, Kansas, Nebraska, New Mexico, Oklahoma, South Dakota, Texas, and Wyoming. The southern portion of the aquifer is the largest source of groundwater for the western interior plains of the United States. Like many aquifers in the West, this once plentiful source of underground water is being depleted rapidly because its water supply is being extracted by thousands of wells at a faster rate than can be replenished through annual rainfall. Falling water tables invariably signal that the withdrawal of groundwater is exceeding the rate of replacement and that, eventually, the source of water could disappear.

Besides the southern Ogallala Aquifer as a groundwater source to some western states, groundwater is available in the southwestern desert basins. The desert basins are the valleys that lie between mountain ranges. Most of these valleys contain groundwater in layers of gravel, sand, silt, and clay, although groundwater conditions vary because of the complex geography and geology of California and other western states. Some of this groundwater is used for irrigation, public supply, and private wells.

Desert basin aquifers are recharged by rainfall and snowmelt runoff from the mountains. However, the groundwater levels are also affected by short- and long-term climatic conditions and by groundwater withdrawals, irrigation returns, and other factors. Complicating the situation is the climate: the arid and semiarid climate found in much of the West results in slow natural replenishment of aquifer water. Slow replenishment coupled with large-scale removal of groundwater results in loss of springs, streams, wetlands, and their associated habitats; degradation of water quality; and land subsidence.

Land subsidence is the sinking of the ground surface caused by the slow drainage of water from the clay and silt sediments in and next to aquifers. As water levels in aquifers decline and the water is drained from the soil, it compacts, causing the land surface to drop. Land subsidence can cause large cracks and holes in the ground, resulting in damage to roads, pipelines, buildings, canals and drainage ditches, railroads, and other structures. The U.S. Geological Survey reports in "Land Subsidence from Ground-Water Pumping" (January 6, 2004, http://geochange.er.usgs.gov/sw/changes/anthropogenic/subside/) that significant land subsidence is occurring throughout the West, such as in Mendota (twenty-nine feet in depth) and Santa Clara Valley, California (twelve feet); Eloy, Arizona (fifteen feet); Houston, Texas (nine feet); and Las Vegas, Nevada (six feet).

DROUGHT
What Is Drought?

Drought is a deficiency of precipitation over an extended period, typically a season or more, and is usually judged relative to a long-term average condition in a particular area. Drought is also related to the timing and the effectiveness of the precipitation. Timing refers to factors such as the period when drought is most likely to occur, delays in the start of the rainy season, and the occurrence of rain in relation to principal crop growth. Precipitation effectiveness refers to the duration, intensity, and frequency of rains or other precipitation events. In many regions of the United States and the world, high temperature, high winds, and low relative humidity are also associated with drought, increasing its severity. In normally arid and semiarid regions, drought refers to dryness over and above the conditions of usual dry seasons.

Figure 8.5 shows drought severity across the forty-eight contiguous states for the week ending February 10, 2007. The map shows that it was extremely moist in the Northeast and areas just south of the Great Lakes. The rest of the eastern portion of the country was near normal in precipitation levels except for southern Florida, which was experiencing moderate drought conditions. Likewise, the central portion of the country was at or above normal precipitation levels except for northern portions of North Dakota, Minnesota, and Wisconsin, which were experiencing moderate to extreme drought. In the West much of Wyoming was in moderate to extreme drought. All of Arizona, most of California, and the southern tip of Nevada were experiencing moderate to severe drought conditions as well.

The interaction between drought, a natural event, and the demand that people place on water supply can worsen a drought's impact. Changes in land use, land degradation, and the construction of dams all affect the characteristics of a water basin (the land area drained by a particular river and its tributaries), and as a result, the pattern and volume of river or stream flow. For example, changes in land use upstream may alter the rates at which water filters into the ground or runs off the ground,

FIGURE 8.5

Drought conditions across the lower 48 states, February 2007

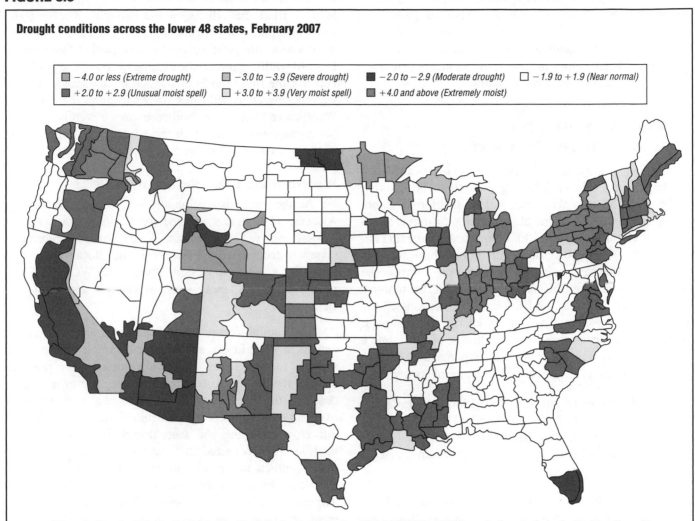

SOURCE: "Drought Severity Index by Division: Weekly Value for Period Ending 10 Feb 2007," in *2007 Drought Severity Index by Division (Long-Term Palmer) Archive*, National Oceanic and Atmospheric Administration, Climate Prediction Center, http://www.cpc.ncep.noaa.gov/products/analysis_ monitoring/regional_monitoring/palmer/2007/02-10-2007.gif (accessed February 15, 2007)

causing more variable stream flow and a higher frequency of water shortage downstream.

Predicting Drought

Anyone can predict with absolute certainty that drought will occur somewhere on the planet, because inevitably it will. It is the how, when, where, and for what duration that are difficult to predict. Drought is never the result of a single cause, but comes from the interaction, and sometimes compounding, of the effects of many causes. On the largest scale, global weather systems play an important part in explaining global and regional weather patterns. These patterns occur with enough frequency and similar characteristics over a sufficient length of time to provide opportunities to strengthen scientists' ability to predict long-range climate, particularly in the tropics. An example of these global systems is El Niño, a disruption of the ocean-atmosphere system in the tropical Pacific Ocean.

On a lesser scale, high-pressure systems inhibit cloud formation and result in lower relative humidity and less precipitation. Regions that are under the influence of high-pressure systems most of the year are generally deserts such as the Sahara and Kalahari deserts in Africa. Most climatic regions experience high-pressure systems at some time, often depending on the season. Prolonged droughts occur when the large-scale deviations in atmospheric circulation patterns persist for months, seasons, or years.

The USGCRP is actively working to promote understanding of climate change and the implications it has for the United States. In *Climate Change Impacts on the United States* (2001, http://www.usgcrp.gov/usgcrp/Library/nationalassessment/ foundation.htm), the USGCRP states that advances in climate science are paving the way for scientists to project climate changes at the regional scale, allowing them to identify regional vulnerabilities and to assess potential regional effects. For example, the USGCRP suggests that the earth's climate has changed in the past and that even greater climate change is likely to occur during the twenty-first century. It also suggests that reduced summer runoff, increased winter runoff, and increased water demands are likely to compound current stresses on water supplies and flood management, particularly in the western United States.

DROUGHT MANAGEMENT
California and the 1986–93 Drought

California is located in a climatic high-pressure zone that hovers off its coast, causing rainfall to be diverted northward. The rain that does fall in California is not evenly distributed. In general, according to the Western Regional Climate Center (2006, http://www.wrcc.dri.edu/pcpn/westus_precip.gif), the northern one-third receives forty to one hundred inches of rain per year, whereas the southern two-thirds of the state receives from five to thirty inches of rain per year, and far southern portions of the state receive less than five inches each year. Although Northern California provides two-thirds of the state's water supply, two-thirds of the state's population lives in Southern California, which has little water of its own.

Drought plagued California from 1986 until 1993, the longest dry period in nearly one hundred years of record keeping. Water supplies dwindled and water rationing was instituted. Many asked the governor to proclaim a state of emergency.

In the San Joaquin Valley, an area the size of Connecticut, the earth dropped more than a foot from land subsidence, damaging roads and buildings. Water resource authorities suspended the agricultural deliveries of water. Crops such as grapevines and fruit trees died because of insufficient water. Farmers who planted water-intensive crops such as cotton, alfalfa, and rice were the hardest hit. Grocery shoppers in every part of the country paid higher prices for some fruits and vegetables because California is the nation's largest agricultural producer.

Throughout the drought, many cities and towns in California instituted severe penalties for excessive water use. Water conservation efforts included installing low-flow showerheads and toilets; not washing cars or filling swimming pools and hot tubs; using dishwashers less frequently, and letting lawns become brown. Water was categorized either as clear (direct from the tap), gray (recycled water from showers, bathtubs, sinks, and washing machines), or black (toilet wastewater). The gray water was reused to water vegetable gardens or plants. Some Californians switched to paper plates to avoid using dishwashers. Others stopped letting the tap water run while brushing their teeth or did not wait for hot water when taking showers. As water rationing became increasingly serious, code enforcers watched for violators and issued citations with fines.

The area's wildlife and vegetation required years to recover from the effects of the drought. The drought years also had a significant impact on trees. Because of the dryness, fire officials continually battled forest fires. In 1990 wildfires forced the closing of Yosemite National Park for the first time in history. Many expensive homes were destroyed as wildfires roared through the canyons. With water from rivers and reservoirs severely limited, helicopters were fitted with large buckets to allow them to scoop water from swimming pools, if necessary.

Western Water Policy Review Act of 1992

At the recommendation of the Western Governors' Association, Congress adopted the Western Water Policy Review Act of 1992, which directed a comprehensive review of federal activities affecting the allocation and use of water in nineteen western states. The Western Water Policy Review Advisory Commission was appointed and chartered in 1995.

The commission released its findings and recommendations in 1997. In the arid West, providing adequate water supplies to meet future demands remained a top priority. Besides the need for more supplies to meet growing water demands, the commission recognized that a need existed to overhaul existing water infrastructure (irrigation canals and ditches, water piping, and water storage devices). The commission also recognized that there were legal and institutional conflicts that needed to be addressed at the federal-state level, between states, and among various water users. The commission recommended the development and implementation of an integrated, coordinated federal policy for federal activities affecting the allocation and use of water in the nineteen western states. The policy was to be developed with the full involvement of the affected states.

National Drought Policy Act of 1998

As a result of the Western Water Policy Review Advisory Commission's report, Congress passed the National Drought Policy Act of 1998. The new law established a National Drought Policy Commission to make recommendations concerning the creation and development of an integrated, coordinated federal drought policy. The commission was to seek public input on recommendations for legislative and administrative actions to help prepare for and alleviate drought's adverse economic, social, health, and environmental effects.

In May 2000 the commission released the report *Preparing for Drought in the 21st Century* (http://govinfo.library.unt.edu/drought/finalreport/fullreport/pdf/reportfull.pdf). The commission recommended the following national policy:

> National drought policy should use the resources of the federal government to support but not supplant nor interfere with state, tribal, regional, local, and individual

efforts to reduce drought impacts. The guiding principles of national drought policy should be:

1. Favor preparedness over insurance, insurance over relief, and incentives over regulation.

2. Set research priorities based on the potential of the research results to reduce drought impacts.

3. Coordinate the delivery of federal services through cooperation and collaboration with nonfederal entities.

The approach of this policy was a marked shift from emphasis on drought relief to that of proactive stance of working to reduce the effects of drought. The commission stated that preparedness was the key to successful drought management and that information and research were needed to support and achieve preparedness. In addition, the commission recommended that the federal government should develop a national drought policy with preparedness at its core and that federal resources should be dedicated to this goal. The commission provided specific recommendations as to how this should be done and urged Congress to pass a National Drought Preparedness Act to achieve the implementation of the recommended policy.

National Drought Preparedness Acts were introduced into Congress in 2002, 2003, and 2005 to establish a National Drought Council within the U.S. Department of Agriculture and to improve national drought preparedness, mitigation, and response efforts. None of these bills became law.

Environmental Quality Incentive Program

Even though a National Drought Preparedness Act has not been passed, the federal government does play a role in assisting farmers to implement technologies and practices to conserve water and to lessen the long-term effects of drought. The Department of Agriculture's Environmental Quality Incentive Program (EQIP) is one such effort. Reauthorized in the Farm Security and Rural Investment Act of 2002, EQIP provides payments to the states to implement conservation practices, paying up to 75% of the cost of certain conservation practices. The National Resources Conservation Service indicates in "National Priorities Programmatic, Fiscal Year 2006" (January 24, 2006, http://www.nrcs.usda.gov/PROGRAMS/natprgmpriorities/FY2006/natprgmprior.html) that one of the national priorities of EQIP is "promotion of conservation of ground and surface water resources." Figure 8.6 shows that a large share of EQIP dollars goes to western states. Table 8.3 shows the exact amounts.

Water Policies—States Lead the Way

Water shortfalls are first and foremost a local and regional problem. Because of the lack of a cohesive federal water policy, states have become important innovators in devising ways to reduce long-term vulnerability to drought. During a widespread drought from 1976 to 1977, no state had a drought plan. In 1982 only three states had them. According to the National Drought Mitigation Center (January 11, 2007, http://drought.unl.edu/plan/stateplans.htm), by 2006 forty-one states had drought plans.

Most state plans do not meet all the goals of the National Drought Mitigation Center recommended planning process. Twenty-nine of the plans address response to droughts rather than mitigation (lessening) of droughts, defining the basic linkages between local, state, and federal entities for coordinated planning and response efforts.

HISTORY OF WATER RIGHTS IN THE WEST

With water scarce in many parts of the arid and semiarid western United States, finding water, bringing it to where it was needed, and obtaining rights to use the water were extremely important aspects of survival. Two important events in the process of settling the West led to laws for the allocation of the scarce water supplies: the discovery of gold and silver in the western mountain regions and the widespread use of irrigation for crop production.

Miners searching for gold and silver diverted stream water into pipes. As a result, an informal code of water regulations started in the mining camps. The first person to file a claim to a gold or silver mine was allowed priority in getting water over any later claims. To remain the owner of a mining claim, the individual had to mark it off, take possession of it, and work the claim productively. This informal water law, conceived more than one hundred years ago, was called the prior appropriation doctrine.

A few years later, this legal practice was adopted by farmers, who needed water for irrigation. The "first in time, first in right" priority system gave the first farmers guaranteed water supplies in times of drought, which were frequent. This right to use water by both the miners and the farmers, who were the first nonnative settlers of the West, was exclusive and absolute. However, the prior appropriation and "first in time, first in right" practices used in the West were different from the system of riparian rights (the right to use water, such as a stream or lake, that abuts one's property) used in the East. Riparian rights could not be sold or transferred, whereas water rights governed by the doctrine of prior appropriation could.

As the western population expanded and states began to write down their laws and arrange them into an organized legal system, the rules for water rights and use changed. The concept of beneficial use became the basis for a landowner's rights to water. Beneficial use has two components: the nature or purpose of the use and the efficient or nonwasteful use of water. State constitutions, statutes, or case law may define the beneficial uses of water. The uses may be different in each state, and the definitions of what uses are beneficial may change over time. The right to use water established under state law

FIGURE 8.6

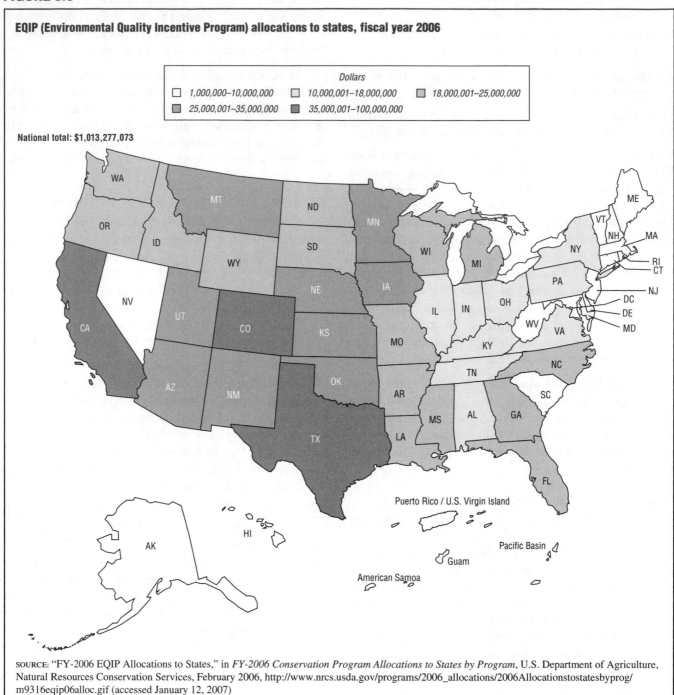

EQIP (Environmental Quality Incentive Program) allocations to states, fiscal year 2006

Dollars

☐ 1,000,000–10,000,000   ☐ 10,000,001–18,000,000   ☐ 18,000,001–25,000,000

☐ 25,000,001–35,000,000   ☐ 35,000,001–100,000,000

National total: $1,013,277,073

SOURCE: "FY-2006 EQIP Allocations to States," in *FY-2006 Conservation Program Allocations to States by Program*, U.S. Department of Agriculture, Natural Resources Conservation Services, February 2006, http://www.nrcs.usda.gov/programs/2006_allocations/2006Allocationstostatesbyprog/ m9316eqip06alloc.gif (accessed January 12, 2007)

may be lost if the beneficial use is discontinued for a prescribed period, frequently summarized as "use it or lose it." Abandonment requires intent to permanently give up the right. Forfeiture results from the failure to use the water in the manner described in state statutes. Either requires a finding by the state resource agency that a water right has been abandoned or forfeited.

Priority determines the order of rank of the rights to use the water in a system—that is, the person first using the water for a beneficial purpose has a right superior to those who begin to use the water at a later date. Priority becomes important when the quantity of available water is insufficient to meet the needs of all those having rights to use water from a common source. Under a priority system, water shortages are not shared as they are under a riparian water rights system. Some western state statutes contain priority or preference categories of water use under which higher-priority uses (such as domestic water supply) have first right to water in times of shortage, regardless of the priority date. There may also be constraints against changes or transfers involving these priority uses.

TABLE 8.3

EQIP (Environmental Quality Incentive Program) allocations in dollars to states, fiscal year 2006

State	Total allocations
Alabama	$16,771,119.00
Alaska	$6,839,921.00
Arizona	$28,328,282.00
Arkansas	$24,604,335.00
California	$62,902,210.00
Colorado	$41,199,573.00
Connecticut	$5,984,300.00
Delaware	$7,618,551.00
Florida	$25,164,310.00
Georgia	$19,050,663.00
Hawaii	$7,510,443.00
Idaho	$20,362,703.00
Illinois	$16,996,755.00
Indiana	$12,956,277.00
Iowa	$25,609,303.00
Kansas	$30,762,396.00
Kentucky	$13,485,727.00
Louisiana	$18,892,373.00
Maine	$0,901,003.00
Maryland	$7,976,993.00
Massachusetts	$5,381,243.00
Michigan	$19,757,606.00
Minnesota	$32,000,245.00
Mississippi	$20,585,619.00
Missouri	$23,389,244.00
Montana	$31,669,601.00
Nebraska	$31,784,747.00
Nevada	$8,256,129.00
New Hampshire	$5,459,975.00
New Jersey	$5,514,610.00
New Mexico	$25,244,238.00
New York	$14,416,790.00
North Carolina	$18,199,282.00
North Dakota	$23,142,407.00
Ohio	$16,855,835.00
Oklahoma	$29,205,352.00
Oregon	$23,901,139.00
Pennsylvania	$13,835,751.00
Rhode Island	$4,878,410.00
South Carolina	$9,794,621.00
South Dakota	$22,003,929.00
Tennessee	$12,841,944.00
Texas	$91,290,491.00
Utah	$25,759,537.00
Vermont	$5,852,213.00
Virginia	$14,981,753.00
Washington	$20,239,965.00
West Virginia	$7,448,115.00
Wisconsin	$20,655,672.00
Wyoming	$18,355,168.00
Pacific basin	$1,610,004.00
Puerto Rico	$6,987,541.00
Total	**$1,013,277,073.00**

SOURCE: "Environmental Quality Incentives Program FY 2006 Financial and Technical Assistance Dollars to States," in *FY-2006 Conservation Program Allocations to States by Program*, U.S. Department of Agriculture, Natural Resources Conservation Services, February 2006, http://www.nrcs.usda.gov/programs/2006_allocations/2006Allocationstostatesbyprog/EQIP06 alloctostates.pdf (accessed January 12, 2007)

Federal Water Laws and Projects

The federal government played a role in both encouraging the economic growth and settlement of the West and developing water laws and projects. The Reclamation Act of 1902 began many years of federal involvement in constructing and subsidizing water projects in the West.

The act was designed to provide subsidized water for small farmers who owned up to 320 acres. Over the years farmers and corporations have used subsidized water to farm thousands of acres by entering into arrangements in which they lease (but do not own) farms.

Based on the existence of irrigated farmland guaranteed by federal subsidies, the West grew rapidly. Cities sprang up in the deserts, attracting a large array of support industries as people from the East and Midwest moved to the Southwest to enjoy the warm, dry climate, stark beauty, and sunshine.

UNANTICIPATED CONSEQUENCES OF IRRIGATION. Because of the composition of the land in much of the West, irrigation practices created some unforeseen problems. Millions of acres of irrigated land overlie a shallow and impermeable clay layer that is sometimes only a few feet below the surface. Significant changes in the land can be caused by the interaction of irrigation water and the clay. During the irrigation season temperatures in much of the desert fluctuate between 90 and 110° F, and some irrigation water is lost to evaporation. When the water evaporates, it leaves behind chemicals called salts, which accumulate in the soil.

The water that is retained in the soil seeps downward, carrying the salts with it, until it hits the impermeable clay layer. Because the water has nowhere to go, it and the dissolved salts rise back up into the plant root zone. However, the high salt concentration in the water can interfere with crop growth. Generally, high salt concentrations hinder the germination of plants and impede their absorption of nutrients. In some cases high salt concentrations in soils have rendered them unable to grow crops. The salts accumulate in the soil as irrigation continues. Eventually, the salts become visible on the ground.

To stop excessive buildup of the salts in the soil, extra irrigation water is required to flush out the salts, generally into surface drainage or groundwater. In locations where these dissolved substances reach high concentrations, the quality of the groundwater and surface water in the area can be harmed. In severe cases the increased salinity renders groundwater and surface water useless for irrigation or drinking and contributes to degraded aquatic habitats.

Buying and Selling Water Rights

Before the mid-1980s the preferred method of getting water was to develop a new supply. As new supplies became less accessible and environmental regulations made supply development more difficult and expensive, creating mechanisms for voluntary water reallocation by buying and selling water rights became more important. Western cities and industries began looking to the agricultural community for water.

Agriculture has traditionally claimed the lion's share of the West's water supplies. If farmers or ranchers, however, could earn more money selling water to a nearby city than spraying it on their crops or watering their stock, shifting the water from farm to city would be in their economic best interest. If the city is saved from damming a local river to increase supplies or depleting an aquifer, it also benefits the environment.

Advocates of the sale of water rights maintain that a free market will allow for more efficient distribution of a source that is often subsidized and just as often squandered. Conservative politicians favor it because it reduces the federal government's role in developing new water supplies. Liberal politicians also like it because more efficient use of water could benefit the environment by lessening the need for dams, which are often environmentally harmful. Since 1981 and continuing into the twenty-first century, western state legislatures have been slowly changing the old laws dealing with water rights to make water right transfers more flexible.

Opponents to the sale of water rights maintain that the sales are draining the life from small, rural communities and can cause irreparable damage to the environment in the long run as the now waterless land is left to crack, bake, and turn into dust. The farmers and ranchers who have refused to sell their water rights are concerned about not only their own water supplies but also the surrounding weeds, dust, and barren land. Once water rights are sold, the use of the land for farming is over.

Water Banks

Not all water right transfers require that water be shifted permanently away from agriculture. Voluntary market transactions can reallocate water on a temporary, long-term, or permanent basis. A water bank (a clearinghouse between the buyers and sellers of water), acting as a water broker and usually subsidized by the state, can be authorized to spend money to buy water from farmers or other sellers who are willing to temporarily or permanently reduce their own use. The bank then resells the water to drinking water suppliers, farmers, ranchers, and industries that need the water.

In *Analysis of Water Banks in the Western States* (July 2004, http://www.ecy.wa.gov/pubs/0411011.pdf), Peggy Clifford, Clay Landry, and Andrea Larsen-Hayden note that water banks exist in almost every western state and are "emerging as an important management tool to meet growing and changing water demands throughout the United States." Clifford, Landry, and Larsen-Hayden identify California, Arizona, and Idaho as states having water banks with a high level of activity.

DESALINATION—A GROWING WATER SUPPLY SOURCE

According to the fact sheet "Desalination" (2007, http://www.awwa.org/Advocacy/pressroom/Desalination.cfm), the American Water Works Association notes that there are over 15,000 desalination plants operating in 120 countries. These plants convert seawater, brackish water (a mixture of seawater and freshwater), and wastewater to freshwater suitable for a variety of purposes. The World Health Organization, in *Nutrients in Drinking Water* (2005, http://www.who.int/water_sanitation_health/dwq/nutrientsindw.pdf), explains that "about 63% of the capacity [to desalinate water] exists in West Asia and the Middle East." According to the Saudi Arabia Information Resource (March 8, 2007, http://saudinf.com/main/a541.htm), Saudi Arabia is a country heavily invested in this technology; it has twenty-seven desalination plants that provide 70% of the country's drinking water. However, Heather Cooley, Peter H. Gleick, and Gary Wolff report in *Desalination, with a Grain of Salt: A California Perspective* (June 2006, http://www.pacinst.org/reports/desalination/desalination_report.pdf) that worldwide, desalination plants in service in 2005 had the capacity to provide only 0.3% of the freshwater used globally.

According to Cooley, Gleick, and Wolff, 56% of global desalination plants are designed to process seawater. Twenty-four percent can process brackish water, and the remaining desalination capacity worldwide is used to desalinate other kinds of water, such as wastewater.

In the United States desalination has become a rapidly growing alternative to water scarcity. With population growth and the threat of drought throughout the United States—particularly in the western states and Florida—desalination, once considered too expensive, is looking more attractive. In fact, Cooley, Gleick, and Wolff indicate that in 2005 the United States was second only to Saudi Arabia in its desalination capacity.

The growth in the use of desalination was fueled by the adoption of the Reclamation Wastewater and Groundwater Study and Facilities Act of 1992. The act directed the secretary of the interior to undertake a program to investigate and identify opportunities for water reclamation and reuse and authorized participation in five water-recycling projects. In 1996 Congress reauthorized the act, expanding it to include another eighteen projects, eight of which are in Southern California, an area in desperate need of water. At the same time, Congress enacted the Water Desalination Act of 1996. The act is based on the fundamental need to find additional sources of potable (safe to drink) water. Its primary goal is development of more cost-effective and technologically efficient means to desalinate water.

Cooley, Gleick, and Wolff state that by January 2005 over two thousand desalination plants had been installed or contracted to be installed in the United States. About 51% of this desalination capacity was used to process brackish water and 26% to process river water. Only 7% of the U.S. desalination capacity was used to desalinate seawater.

Desalination Processes

Desalination is the removal of dissolved minerals (including, but not limited to, salts) from seawater, brackish water, or treated wastewater. A number of technologies have been developed for desalination. In the United States desalination research is directed by the Bureau of Reclamation, which is a branch of the U.S. Department of the Interior.

There are several desalination processes:

- Reverse osmosis—filtered water is pumped at high pressure through permeable membranes, separating the salts from the water.

- Distillation—water is heated and then evaporated to separate out the dissolved minerals. The most common methods of distillation are:

1. Multistage flash distillation, in which the water is heated and the pressure lowered so that the water flashes into vapor that is drawn off and cooled to provide desalted water.

2. Multiple effect distillation, in which the water passes through a number of evaporators in series with the vapor from one series being used to evaporate the water in the next series.

3. Vapor compression, in which the water is evaporated and the vapor compressed; the heated compressed vapor is used to evaporate additional water.

4. Electrodialysis—electric current is applied to brackish water, causing positive and negative ions of dissolved salt to split apart.

The two most common desalination processes worldwide are multistage flash distillation and reverse osmosis. Although water of different quality, including seawater, brackish water, or impure industrial wastewater, can be desalinated, seawater and brackish water are the most common water sources worldwide.

Desalination Plants in the United States

Buckeye, Arizona, became the first town in the United States to have all its water supplied by its own electrodialysis-desalting plant. Starting in 1962, the plant provided about 650,000 gallons of water daily at a cost of about $1 per 1,670 gallons. In 1967 Key West, Florida, opened a flash-evaporation plant and became the first city in the United States to draw its freshwater from the sea. Cooley, Gleick, and Wolff note that as of January 2005 the states having desalination capacity of more than 1% of the total U.S. capacity were, from most to least: Florida, California, Arizona, Texas, Virginia, Colorado, Pennsylvania, Ohio, Alabama, North Carolina, Utah, Oklahoma, and Hawaii.

According to the Bureau of Reclamation's Yuma Area Office (March 2007, http://www.usbr.gov/lc/yuma/), the Yuma desalting plant in Arizona is the largest reverse osmosis desalting plant in the world. The plant was built as a result of a dispute with Mexico over the salinity of drainage water from the Wellton-Mohawk Irrigation District in Arizona. The salinity of this irrigation return flow caused marked deterioration in Colorado River water quality in Mexico. The problem was so severe that Congress enacted the Colorado River Basin Salinity Control Act to fund the plant's construction.

This is how the plant works: The saline drainage water from farmlands east of Yuma flows in a concrete drainage canal to the desalting plant. The drainage water enters the plant intake system, where screens remove algae and large debris such as tree limbs. As the drainage water flows into the plant, it is treated with chlorine to kill organisms and stop the growth of algae, which would damage or plug the filters and membranes. Before being desalted, the water passes through several pretreatment steps to remove all solids that would interfere with the membrane performance. Pretreatment extends the life of the reverse osmosis membranes three to five years. Without pretreatment, the membranes would last about one hour. It takes only a few hours for a unit of water to travel through the plant, from where it enters as untreated drainage water, is pretreated and subjected to reverse osmosis, and then discharged to a small canal that empties into the Colorado River.

Cooley, Gleick, and Wolff describe how California is proposing an expansion in its desalination capacity that would increase its seawater desalination capacity by seventyfold. Figure 8.7 shows the locations of the proposed desalination plants. Eleven of the plants were proposed for Northern California and ten for Southern California. Of the twenty-one proposed plants, twelve are slated to be larger than any previous desalting plants built in that state. (Note that not all the proposed plants have remained under active consideration.)

Advantages and Disadvantages of Desalination

One of the most important factors determining whether desalination is a viable way to provide water to consumers is cost. Cooley, Gleick, and Wolff calculate that desalination of water in California will cost users between $3 and $10 per thousand gallons. There are many reasons, such as subsidies and energy costs, for the large variation in the price, but one major reason is the cost of water distribution in various regions of the state. This price range is far above prices consumers generally pay for water. Farmers may pay as little as $0.20 to $0.40 per thousand gallons, whereas consumers living in urban areas may pay $1 to $3 per thousand gallons. Clearly, desalinated water is expensive, and while the long-term trend in desalination costs has been downward, in recent years prices have increased because

FIGURE 8.7

Proposed desalination plants in California, 2006

● >20 MGD (76,000 m³/d)
◉ 5–20 MGD (19,000–76,000 m³/d)
● <5 MGD (19,000 m³/d)

Notes: MGD=million gallons per day.
m³/d:=cubic meters per day.

SOURCE: Heather Cooley, Peter H. Gleick, and Gary Wolff, "Figure 13. Map of Proposed Desalination Plants in California as of Spring 2006," in *Desalination, With a Grain of Salt: A California Perspective*, Pacific Institute for Studies in Development, Environment, and Security, June 2006, http://www.pacinst.org/reports/desalination/desalination_report .pdf (accessed February 16, 2007)

TABLE 8.4

The five realities of water in the West

Five interrelated realities of water management will shape, if not control, policy level water supply decisions in the West through 2025:

1) Explosive population growth in areas of the West where water is already scarce.
2) Water shortages occur frequently in the West.
3) Over-allocated watersheds can cause crisis and conflict.
4) Water facilities are aging.
5) Crisis management is not effective in dealing with water conflicts.

SOURCE: Adapted from "Water 2025 Realities," in *Water 2025: Preventing Crises and Conflict in the West, Water 2025 Status Report* U.S. Department of the Interior, Bureau of Reclamation, August 2005, http://www.doi.gov/ water2025/Water%202025-08-05.pdf (accessed January 11, 2007)

can be reused for irrigation and industrial use and to maintain stream flow.

Indirect reuse of treated municipal wastewater (reclaimed water) is becoming increasingly attractive to many municipalities, especially in the West. For example, the Orange County Water District in California has implemented the Green Acres Project, which uses reclaimed water for landscape irrigation at parks and schools, as well as for various industrial uses. According to the district (October 31, 2001, http://www.ocwd.com/_html/gap.htm), the project can purify up to 7.5 million gallons per day of reclaimed water from the Orange County Sanitation District. Using a newly built reservoir, the project can store up to 350 million gallons of this recycled water per year.

WATER 2025

Problems in the West, including explosive population growth, existing water shortages, conflicts over water, aging water facilities, and ineffective crisis management, have led to a Interior Department proposal designed to assist communities in addressing these needs. In *Water 2025: Preventing Crises and Conflict in the West* (August 2005, http://www.doi.gov/water2025/Water%202025-08-05.pdf), the Interior Department calls for concentrating existing federal financial and technical resources in key western watersheds and in critical research and development, such as water conservation and desalinization, that will help predict, prevent, and alleviate water supply conflicts. Table 8.4 outlines the five realities of water in the West as identified in the report.

Water 2025 emphasizes the need for states, tribes, local governments, and the public to decide how best to resolve the water supply crisis in the West. As part of this plan, the Interior Department prepared an analysis of potential water supply crises and conflicts that may occur by 2025. The department intended to seek extensive input from states, tribes, and the public on the prepared analysis to revise and improve the analysis as needed.

of rising energy and construction costs. In the early 2000s cost estimates for new plants were noticeably higher than those for similar plants built just a few years earlier

Nonetheless, desalination can provide a reliable source of water independent of the weather and can provide an additional source of water to those already in place. In addition, the desalination process removes water impurities. However, water quality must be monitored to ensure than contaminants are not added during the desalination process or that essential minerals are not removed.

WATER REUSE

Wastewater from sewage treatment plants is one of the largest potential sources of freshwater where supplies are limited. After it has been treated to kill pathogens (disease-causing organisms) and remove contaminants, it

FIGURE 8.8

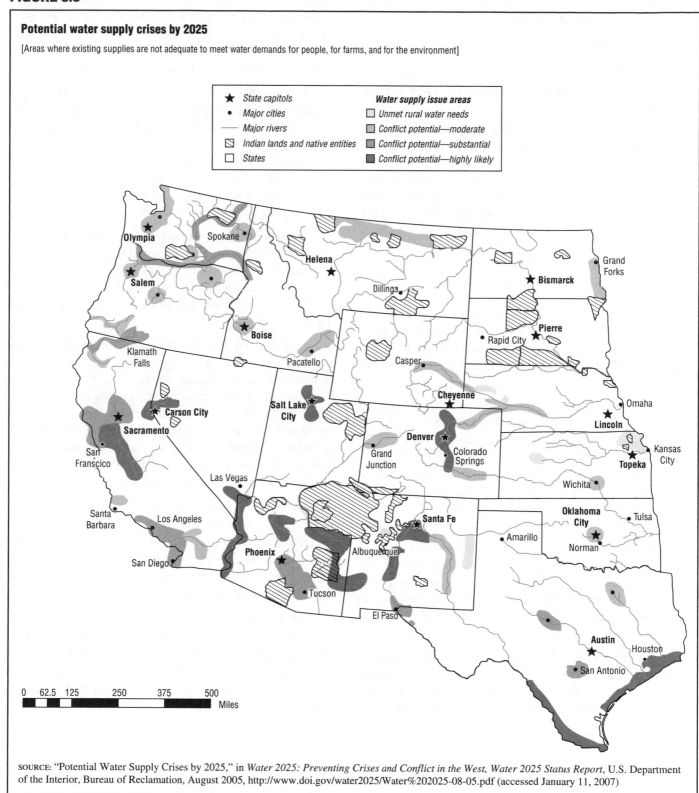

Potential water supply crises by 2025

[Areas where existing supplies are not adequate to meet water demands for people, for farms, and for the environment]

★ State capitols
• Major cities
— Major rivers
▨ Indian lands and native entities
☐ States

Water supply issue areas
☐ Unmet rural water needs
▨ Conflict potential—moderate
▨ Conflict potential—substantial
■ Conflict potential—highly likely

SOURCE: "Potential Water Supply Crises by 2025," in *Water 2025: Preventing Crises and Conflict in the West, Water 2025 Status Report*, U.S. Department of the Interior, Bureau of Reclamation, August 2005, http://www.doi.gov/water2025/Water%202025-08-05.pdf (accessed January 11, 2007)

Water 2025 is a departure from previous plans in that it focuses on strategies and measures that can be put in place before events such as drought will bring further divisiveness to communities in the West. The department believes that conflict can be minimized or avoided when potential water supply crises are addressed in advance by local and regional communities. Figure 8.8 shows areas of potential water supply crises by 2025. The four key tools of the *Water 2025* water-crisis prevention efforts are:

- Water conservation, water-use efficiency, and the use of market-based approaches

- Collaboration

- Improved technology

- Removal of institutional barriers and increase in inter-agency coordination

WATER CONSERVATION, WATER-USE EFFICIENCY, AND MARKETS. *Water 2025* identifies supervisory control and data acquisition systems as one area in which water conservation efforts can be improved in the management of rivers. These systems allow river managers to remotely monitor and operate key river and canal facilities on a real-time basis. The Interior Department recommends that individual stations be set to monitor river levels or flow rates continuously. This will help the Interior Department and water district managers respond to daily water management needs and emergencies in a timely fashion by controlling pump and canal facilities remotely. The Interior Department notes that even though the cost of this high-tech equipment has become more affordable over time, less than 20% of irrigation water delivery systems use this technology.

Water 2025 notes that for every dollar spent on canal modernization (such as rehabilitating canal gates), an expected return of $3 to $5 in conserved water can be achieved. In addition, for every dollar spent on maintaining an existing canal lining, a return of up to $10 in conserved water can be achieved. Canal-lining technologies have reduced seepage losses in central Oregon by as much as 50%.

Additional measures recommended by *Water 2025* include improvement in design and construction of new measuring devices for irrigation water delivery systems, continued support for water banks and water markets, and interagency efforts to coordinate existing and new water conservation programs.

COLLABORATION. *Water 2025* points to litigation over competing water rights as one of the problems affecting water supply and conservation efforts in the West. Water managers sometimes must wait years or even decades until adjudication (consideration of the facts and rendering a decision) is completed. In the meantime they do not know how to allocate water in times of scarcity. The Interior Department is finding ways to accelerate court proceedings to protect existing federal and nonfederal rights.

An example of successful facilitation efforts occurred in California. According to the Bureau of Reclamation, for more than two decades the East Bay Municipal Utility District and several localities struggled over the management of the Sacramento River, resulting in the disruption of the efficient use of water. Facilitation sponsored by the bureau led to a sustainable and locally developed agreement among the interested parties.

IMPROVED TECHNOLOGY. Recognizing that wastewater, saltwater, and other impaired water can be purified to increase their usefulness, the Interior Department is researching ways to reduce the high costs that impede the adoption of new water purification technologies including desalination.

REMOVAL OF INSTITUTIONAL BARRIERS AND INTER-AGENCY COOPERATION. According to the Interior Department, in some areas of the West federal facilities have excess capacity during certain times of the year that could be used to satisfy unmet demands elsewhere. Sometimes this excess capacity is not available because of policy or legal constraints. The department believes that in some cases this additional capacity can be made available with appropriate changes in its policies.

The Interior Department is cooperating with other federal agencies to more effectively focus federal dollars on critical water shortage areas. Through active support of the National Drought Monitoring Network the department is helping accelerate the development of strategies for drought preparedness.

Additional measures include the formation of drought action teams to focus scarce resources quickly when and where they are needed, and the publication by the Geological Survey of water resource assessments online so that decision makers can better understand the water supply component of drought conditions.

IMPORTANT NAMES
AND ADDRESSES

American Ground Water Trust
16 Centre St.
Concord, NH 03301
(603) 228-5444
FAX: (603) 228-6557
URL: http://www.agwt.org/

American Petroleum Institute
1220 L St. NW
Washington, DC 20005-4070
(202) 682-8000
URL: http://www.api.org/

American Water Works Association
6666 W. Quincy Ave.
Denver, CO 80235
(303) 794-7711
1-800-926-7337
FAX: (303) 347-0804
URL: http://www.awwa.org/

Association of State Drinking Water Administrators
1401 Wilson Blvd., Ste. 1225
Arlington, VA 22209
(703) 812-9505
FAX: (703) 812-9506
URL: http://www.asdwa.org/

Bureau of Land Management
Office of Public Affairs
1849 C St., Rm. 406-LS
Washington, DC 20240
(202) 452-5125
FAX: (202) 452-5124
URL: http://www.blm.gov/

Centers for Disease Control and Prevention
1600 Clifton Rd.
Atlanta, GA 30333
(404) 639-3311
1-800-311-3435
URL: http://www.cdc.gov/

Chesapeake Bay Program
410 Severn Ave., Ste. 109
Annapolis, MD 21403
1-800-YOUR BAY
FAX: (410) 267-5777
URL: http://www.chesapeakebay.net/

Ducks Unlimited
One Waterfowl Way
Memphis, TN 38120
(901) 758-3825
1-800-45-DUCKS
FAX: (901) 758-3850
URL: http://www.ducks.org/

Environmental Protection Agency
Ariel Rios Bldg.
1200 Pennsylvania Ave. NW
Washington, DC 20460
(202) 272-0167
FAX: (202) 501-1450
URL: http://www.epa.gov/

International Bottled Water Association
1700 Diagonal Rd., Ste. 650
Alexandria, VA 22314
(703) 683-5213
FAX: (703) 683-4074
URL: http://www.bottledwater.org/

National Drought Mitigation Center
University of Nebraska at Lincoln
819 Hardin Hall
3310 Holdrege St.
Lincoln, NE 68583-0988
(402) 472-6707
FAX: (402) 472-2946
E-mail: ndmc@drought.unl.edu
URL: http://www.drought.unl.edu/

National Oceanic and Atmospheric Administration
Fourteenth St. and Constitution Ave.
NW, Rm. 6217
Washington, DC 20230

(202) 482-6090
FAX: (202) 482-3154
URL: http://www.noaa.gov/

Natural Resources Conservation Service
Fourteenth and Independence Ave. SW
Washington, DC 20250
URL: http://www.nrcs.usda.gov/

National Wildlife Federation
11100 Wildlife Center Dr.
Reston, VA 20190
1-800-822-9919
(703) 438-6000
URL: http://www.nwf.org/

Nature Conservancy
4245 N. Fairfax Dr., Ste. 100
Arlington, VA 22203-1606
(703) 841-5300
FAX: (703) 841-1283
URL: http://www.nature.org/

Soil and Water Conservation Society
945 SW Ankeny Rd.
Ankeny, IA 50023
(515) 289-2331
FAX: (515) 289-1227
URL: http://www.swcs.org/

U.S. Fish and Wildlife Service
1849 C St. NW
Washington, DC 20242
(202) 208-5634
1-800-344-WILD
URL: http://www.fws.gov/

U.S. Geological Survey
12201 Sunrise Valley Dr.
Reston, VA 20192
(703) 648-4000
FAX: (703) 648-5548
URL: http://www.usgs.gov/

U.S. Government Accountability Office
441 G St. NW
Washington, DC 20548
(202) 512-3000
FAX: (202) 512-8546
URL: http://www.gao.gov/

U.S. Water News
230 Main St.
Halstead, KS 67056
(316) 835-2222
FAX: (316) 835-2223
URL: http://www.uswaternews.com/

Water Environment Federation
601 Wythe St.
Alexandria, VA 22314-1994
1-800-666-0206
FAX: (703) 684-2492
URL: http://www.wef.org/

RESOURCES

Responsibility for the protection, management, and use of water is spread across many federal agencies. The U.S. Geological Survey (USGS), a branch of the U.S. Department of the Interior, has the principal responsibility within the federal government for appraising the nation's resources and providing hydrological information. The USGS publications consulted for this edition include *Ground Water and Surface Water: A Single Resource* (Thomas C. Winter et al.; 1998), *Estimated Use of Water in the United States in 2000* (Susan S. Hutson et al., 2004), "Restoring Life to the Dead Zone: Addressing Gulf Hypoxia, a National Problem" (June 2000), *Where Is Earth's Water Located?* (August 2006), *Analysis of Water Use in the Piscataqua River and Coastal Watersheds, Southeastern New Hampshire* (2006), "Groundwater" (2006), and *The Quality of Our Nation's Waters: Volatile Organic Compounds in the Nation's Ground Water and Drinking-Water Supply Wells* (2006).

The U.S. Environmental Protection Agency (EPA) is the federal regulatory agency charged with, among other things, protection of both surface and groundwater quality, overseeing the states' management of drinking water protection programs, the development of water quality standards, and the enforcement of laws addressing water quality. The EPA published the *2000 National Water Quality Inventory* (August 2002), *The National Coastal Condition Report II (2005)* (December 2004), and *The Wadeable Streams Assessment: A Collaborative Survey of the Nation's Streams* (December 2006). Other EPA documents used in this book include *EPA National Primary Drinking Water Standards* (June 2003), *Safe Drinking Water Act, Section 1429 Groundwater Report to Congress* (October 1999), *National Water Quality Inventory: 1998 Report to Congress* (June 2000), "Safe Drinking Water Act 30th Anniversary: Water Facts" (June 2004), "Safe Drinking Water Act 30th Anniversary Drinking Water Monitoring, Compliance, and Enforcement" (June 2004), *Cleaning Up Leaks from Underground Storage Tanks* (February 2005),

Factoids: Drinking Water and Ground Water Statistics for 2005 (December 2006), and *Implementing the BEACH Act of 2000: Report to Congress* (October 2006).

The U.S. General Accounting Office (now the U.S. Government Accountability Office), the investigative arm of Congress, issued *Water Pollution: Information on the Use of Alternative Wastewater Treatment Systems* (September 1994), *Hazardous Waste: Compliance with Groundwatering Requirements at Land Disposal Facilities* (1995), *Oregon Watersheds: Many Activities Contribute to Turbidity During Large Storms* (July 1998), *Water Quality: Key EPA and State Decisions Limited by Inconsistent an Incomplete Data* (March 2000), *Water Infrastructure: Water-Efficient Plumbing Fixtures Reduce Water Consumption and Wastewater Flows* (August 2000), *Freshwater Supply: States' Views of How Federal Agencies Could Help Them Meet the Challenges of Expected Shortages* (July 2003), and *Environmental Protection: More Complete Data on Continued Emphasis on Leak Prevention Could Improve EPA's Underground Storage Tank Program* (November 2005).

The National Oceanic and Atmospheric Administration (NOAA), a branch of the U.S. Department of Commerce, has the principal responsibility within the federal government for appraising and protecting the nation's aquatic resources, managing its fisheries, and predicting weather and climate. NOAA documents and publications used for this book include *The Deadliest, Costliest, and Most Intense United States Hurricanes from 1900 to 2000* (Jerry D. Jarrel et al.; October 2001), *Billion Dollar U.S. Weather Disasters* (2006), *Population Trends along the Coastal United States: 1980–2008* (Kristen M. Crossett et al.; September 2004), *The State of Coral Reef Ecosystems of the United States and Pacific Freely Associated States: 2005* (August 2005), and *Drought Severity Index by Division: Weekly Value for Period Ending 10 Feb 2007* (2007).

The U.S. Fish and Wildlife Service (USFWS), a branch of the Department of the Interior, is charged with protection of living resources. USFWS documents used in this edition include *Status and Trends of Wetlands in the Conterminous United States 1998 to 2004* (2006) and the *2001 National Survey of Fishing, Hunting and Wild-Life Associated Recreation* (October 2002).

The U.S. Department of Agriculture (USDA) provides national leadership on natural resources. USDA documents and publications consulted for this edition include *FY-2006 Conservation Program Allocations to States by Program* (2006), *National Resources Inventory 2003 Annual NRI* (2006), and *Past and Future Freshwater Use in the United States* (December 1999).

The Centers for Disease Control and Prevention (CDC) are branches of the U.S. Department of Health and Human Services. CDC documents used in this book include "Surveillance for Waterborne Disease and Outbreaks Associated with Drinking Water and Water Not Intended for Drinking—United States, 2003–2004" (December 2006) and "Surveillance for Waterborne Disease and Outbreaks Associated with Recreational Water—United States, 2003–2004" (December 2006). Both documents are published in the CDC's periodical *Morbidity and Mortality Weekly Update.*

Additional important publications about water include the U.S. Department of the Interior's *Water 2025: Preventing Crises and Conflict in the West* (August 2005), the U.S. Global Change Research Program's *Climate Action Report 2002: The United States of America's Third National Communication under the United Nations Framework Convention on Climate Change* (May 2002), and the United Nation's *Water—A Shared Responsibility: World Water Development Report 2* (March 2006). The Gallup Poll provided Americans' opinions on water issues in two documents: *Water Pollution Tops Americans' Environmental Concerns* (Joseph Carroll; April 2006) and *Gallup's Pulse of Democracy: Environment* (March 2006).

INDEX

Texas
  coral reef of, 105
  estuary program in, 108
  hazardous waste generation, 78
  industrial water withdrawals, 22
  off-stream water use by, 12
  Ogallala Aquifer and, 63
  public-supply water use, 16
  water use for livestock, 22
  water use for mining, 24
  water use for thermoelectric power, 25
Texas Department of Water Resources, 63
Thermal springs, 65
Thermoelectric power
  freshwater usage projections, 30
  off-stream water use for, 14
  water use for, 25
  water use (fresh/saline), by state, 26f
Third world, drinking water in, 97–98
Timing, of precipitation, 140
Toilets, 31–32, 91
Torrey Canyon (supertanker), 113
Total coliform bacteria, 87
Total trihalomethanes (TTHMs), 88
Transient noncommunity water systems, 82, 82t
Transit point, 137
Transpiration, 4
Travel Montana, 65
Treaties. *See* Legislation and international treaties
Treatment train, 88
Trihalomethanes, 73
Trout Unlimited v. U.S. Department of Agriculture, 27
Tsunamis, 103–104
TTHMs (total trihalomethanes), 88
Tulare Lake Basin Water Storage District v. U.S. 49 Fed. Cl. 313, 27
Turbidity
  activities within watershed that can increase, 86f
  of ocean shoreline waters, 111
  wetlands and, 124
2005 Review: Water Quality Standards for Salinity, Colorado River System (Colorado River Basin Salinity Control Forum), 137
2001 National Survey of Fishing, Hunting and Wildlife-Associated Recreation (USFWS), 125
Type E botulism, 58
Typhoid fever, 85, 88
Typhoon, 103

U

Ultraviolet, 88
Unconfined aquifer, 62
Underground Injection Control (UIC) Program, 78

Underground Storage Tank (UST) Program, 75–77
Underground storage tanks (USTs)
  cleanup of, 76(f4.11)
  groundwater contamination from leaking underground storage tanks, 76(f4.10)
  as groundwater contamination source, 74–77
United Nations (UN)
  on access to safe water, 97
  publication on water, 154
  on water shortages, 8–9
United States
  amount of water used by Americans, 81
  coral reef ecosystems in U.S. and FAS, threat levels to, 106f
  coral reefs in, 105, 105f, 107
  Great Lakes Water Quality Agreement, 57–58
  hurricanes, costliest, 104t
  hurricanes, deadliest mainland U.S., 103(t6.3)
  water use, future of, 29–34
  water use trends since 1950, 28–29, 29t
United States Army Corps of Engineers, Solid Waste Agency of Northern Cook County v., 129
University of North Carolina at Chapel Hill, 81
Unsaturated zone, 61
Urban runoff, water pollution from, 7, 7f, 7t
Urbanization, 8, 128
U.S. Census Bureau, 34, 133
U.S. Coast Guard, 113, 114
U.S. Department of Agriculture, Trout Unlimited v., 27
U.S. Department of Agriculture (USDA)
  Environmental Quality Incentive Program, 143
  on Klamath Basin, 126
  publications of, 154
  water quality management and, 40
U.S. Department of the Interior
  desalination research by, 147
  Water 2025: Preventing Crises and Conflict in the West, 148–150, 154
U.S. Environmental Protection Agency (EPA)
  algal blooms and, 118
  beaches and, 111–112
  on constructed treatment wetlands, 131
  on coral reefs, 105
  on cruise ship waste, 115
  drinking water regulations by year enacted, 90t
  drinking water requirements, 87
  on estuaries, 107–108
  Federal Water Pollution Control Act and, 91–92
  fish advisories, 58

on groundwater contamination, 69, 70, 74, 78, 79
on groundwater quality, 67–68
on High Plains Aquifer, 63
on household water use, 18
on lakes, 38
Lead and Copper Rule, 85
on nearshore waters, 107
on oil spills, 113
on private water systems, 83
on public-supply water systems, 81–82
publications of, 153
recreational water-associated outbreaks, 58–60
Safe Drinking Water Act and, 90–91
on salinity of ocean, 101–102
on salmon recovery project, 28
on SDWA violations, 92–93
underground storage tanks and, 75–76
on water conservation, 33
on water in food, 1
water quality, assessment/monitoring of, 39–40
on water quality of lakes, 51–56
water quality of nation's waterways, 110–111
on water quality of streams, 40–41, 43–51
water safety and bioterrorism, 36
on waterborne-disease outbreaks, 93–94
on wetlands, 119, 122, 126
U.S. Fish and Wildlife Service (USFWS)
  exotic species regulation, 118
  on Klamath Basin, 126
  publications of, 154
  on water rights, 27, 28
  on waterfowl, 123
  on wetland-related recreation, 125
U.S. Food and Drug Administration, 96
U.S. Forest Service, 27
U.S. 49 Fed. Cl. 313, Tulare Lake Basin Water Storage District v., 27
U.S. General Accounting Office. *See* U.S. Government Accountability Office
U.S. Geological Survey (USGS)
  on amount of groundwater, 61
  on drinking water suppliers, 81
  on evapotranspiration, 4
  on freshwater supply, 6, 11
  groundwater contamination and, 69–70
  on groundwater uses, 65
  on High Plains Aquifer, 63
  on hydrologic cycle, 2
  on hypoxia in Gulf of Mexico, 116
  on lakes, 39
  on land subsidence, 140
  on largest rivers, 38
  on ocean, 101
  publications of, 153